AUSTENITIC STAINLESS STEELS - NEW ASPECTS

Edited by **Wojciech Borek, Tomasz Tański**
and **Zbigniew Brytan**

Austenitic Stainless Steels - New Aspects
http://dx.doi.org/10.5772/67935
Edited by Wojciech Borek, Tomasz Tański and Zbigniew Brytan

Contributors

Javad Mola, Roland Loto, C Shanjeevi, Lev Pisarevskiy, Georgiy Filippov, Maribel Saucedo-Muñoz, Grzegorz Golański, Adam Zieliński, Hanna Purzyńska, Jerzy Kaleta, Przemysław Wiewiórski, Wojciech Wiśniewski, Wojciech Borek, Tamazi Eterashvili, Ladislav Kander, Zdeněk Kuboň

Published by InTech
Janeza Trdine 9, 51000 Rijeka, Croatia

Publishing Process Manager Maja Bozicevic
Technical Editor SPi Global
Cover InTech Design team

First published December, 2017
Printed in Croatia
Legal deposit, Croatia: National and University Library in Zagreb

Additional hard copies can be obtained from orders@intechopen.com

Austenitic Stainless Steels - New Aspects, Edited by Wojciech Borek, Tomasz Tański and Zbigniew Brytan
p. cm.
Print ISBN 978-953-51-3701-6
Online ISBN 978-953-51-3702-3

For EU product safety concerns:
IN TECH d.o.o., Prolaz Marije Krucifikse Kozulić 3, 51000 Rijeka, Croatia,
info@intechopen.com or visit our website at intechopen.com.

Meet the editors

Dr. Wojciech Borek, PhD and MSc in Engineering, is an assistant professor in the Institute of Engineering Materials and Biomaterials at the Silesian University of Technology in Gliwice, Poland, and a member of the association of Polish Cluster of Innovative Forging Technologies "HEFAJSTOS." His scientific interests include materials science, heat treatment, thermomechanical treatment, plastic deformations, and Gleeble simulations; also, he is a specialist in steels, stainless steel, high-manganese austenitic steels and light metal alloys. He is an author and coauthor of ca. 60 scientific publications worldwide including 2 chapters in books and more than 17 publications in the Philadelphia list; he won 10 awards and honors, national and international; has served, or is currently serving, as a contractor of more than 6 research and didactic projects in Poland and abroad and a reviewer of numerous scientific publications and co-promoter of two doctoral dissertations.

Prof. Tomasz Tański is the Head of the Institute of Engineering Materials and Biomaterials Silesian University of Technology; a member of the Metallurgy Section of Committee of Metallurgy of the Polish Academy of Sciences; a specialist in materials such as non-ferrous alloys, composite and nanostructured materials, manufacturing engineering; surface, properties and structures of engineering materials. He authored or co-authored more than 300 scientific publications worldwide including 12 monographs and books, more than 70 publications in the Philadelphia list; won 18 National and International Awards and Honors; has served, or is currently serving, as a supervisor or contractor for more than 15 research and didactic projects in Poland and abroad. Prof. Tański is also a reviewer and promoter of numerous scientific papers, including 4 doctoral research in the field of nanotechnology and materials.

Dr. Zbigniew Brytan, PhD and MSc in Engineering, is an assistant professor in the Institute of Engineering Materials and Biomaterials at the Silesian University of Technology in Gliwice, Poland. His scientific interests include corrosion resistant alloys, stainless steel, sintering, surface engineering, laser processing of metals, welding of stainless steels. He is an author and coauthor of ca. 50 scientific publications worldwide, including 1 book and more than 20 publications in the Philadelphia list. He has served as a main contractor or contractor of more than five research and didactic projects in Poland in the field of stainless steels. He is known as the expert in stainless steel processing, providing technical expertise to more than 30 stainless steel producers. He is also a trainer for product managers, sellers and process engineers of stainless steels.

Contents

Preface

Austenitic stainless steels are the most popular and well known group of stainless steels. Global production of austenitic stainless steel in the last 15 years is still increasing, which indicates that this group of steel has a very bright future and specific research and development in this field are still required. This group of austenitic stainless steel has many advantages, especially corrosion resistance. However, even this group of steel is not 100% safe from corrosion. That is why the right chosen steel needs to be used in proper conditions and appropriate connection must be employed.

This book intends to provide the reader, not only for students but also for professional engineers who are working in the industry as well as to specialists, a comprehensive overview of the state-of-the-art in new trends, research results and development of austenitic stainless steel. Chapters for this book have been written by respected and well-known researchers and specialists from different countries. We hope that after studying this book, you will have objective knowledge about new aspects in the topic concerning austenitic stainless steels.

Wojciech Borek, Tomasz Tański and Zbigniew Brytan
Institute of Engineering Materials and Biomaterials
Silesian University of Technology
Gliwice, Poland

Introductory Chapter: Why Austenitic Stainless Steels are Continuously Interesting for Science?

Zbigniew Brytan, Wojciech Borek and
Tomasz Tański

Additional information is available at the end of the chapter

http://dx.doi.org/10.5772/intechopen.72062

1. About Austenitic Stainless Steels

The history of stainless steels begun in the twentieth century with first works of Monnartz (Germany) in 1908 that discovered passivity in Fe-Cr alloys when Cr content reach 12% and introduced the property of "stainlessness" for metallic alloys. Thanks for studies on Cr-Ni-Fe alloys made by Mauer and Strauss (Germany), the first patents were issued in 1912 for two nickel-containing stainless steels. In this way, the history of austenitic stainless steel began. Today the austenitic stainless steel of classic composition 18%Cr, 8%Ni (grade 304 L) is still the most widely used by far in the world. The first commercial melt of austenitic stainless steels was emerged in 1913 due to works of Hary Brearley (G.B.). In 2013, Stainless Steel celebrated its 100 years and was one of the significant materials' developments in the last century. The stainless steel is still one of the fastest growing materials. The unique characteristic of stainless steel arises from three main factors. The versatility is resulting from high corrosion resistance, excellent low- and high-temperature properties, high toughness, formability, and weldability. The long life of stainless steels has been proven in service in a wide range of environments, together with low maintenance costs compared to other highly alloyed metallic materials. The retained value of stainless steel results from the high intrinsic value (contains expansive alloying elements—nickel) and easy recycling. Stainless steel, especially of austenitic microstructure, plays a crucial role in achieving sustainable development, nowadays so important for further generations [1–3].

The growing consumption of austenitic stainless steels is driven by scientific developments in this field, regarding new grades of improved or optimized properties, studies on corrosion resistance in various environments or microstructural phase transformations ongoing during service or fabrication. The purpose of the book is to present most exciting field of scientific research related to austenitic stainless steels and creep-resistant austenitic alloys. Present chapters deal with different aspects of alloy design. First of all, are associated with the effect of work hardening on microstructure and mechanical properties. One of the most exciting field in austenitic stainless steel development concern studies on work hardening mechanism of Fe-Cr-Mn-Ni-based austenitic stainless steels. Understanding the influence of deformation-induced processes on the strain hardening behavior and tensile elongation is essential to the economical design of highly formable austenitic stainless steels. The occurrence of various deformation-induced processes such as deformation-induced ε/α'-martensite formation mechanisms in austenitic steels is governed by the stacking fault energy (SFE). The influence of alloying elements on the SFE of austenitic stainless steels can be deduced from their influence on the $M_d^{\gamma\rightarrow\alpha'}$ temperature. Therefore, relationships giving the compositional dependence of $Md^{\gamma\rightarrow\alpha'}$ temperature can be used as guidelines for the design of austenitic stainless steels with improved properties [4, 5].

Austenitic stainless steels during austenite to martensite transformation may develop the magnetic response, as martensite is ferromagnetic. Therefore, the martensitic transportation induced by plastic deformation in austenitic stainless steels can also be studied using selected cross effects. The application of the magnetomechanical effect (the Villari effect) and the thermomechanical effect (the Kelvin/Thomson effect) turned out to be particularly useful in this case because, they change significantly with martensite initiation and then accumulation in austenite. This approach gives the opportunity to develop non-destructive methods of investigating the martensite transformation and allows follow and visualizes transformations online during the fatigue process, without the necessity to use, for example, roentgenographic or microscopic methods. This method can also be introduced in both laboratory conditions and on real constructions made of metastable austenitic steels [6, 7].

The book chapters discuses also an essential field of studies concerning microstructural phase transformations ongoing during service at high- and low-temperature conditions. The stainless steels are susceptible to secondary precipitates under service. Precipitation of sigma phase represents one of the most potentially dangerous degradation mechanisms in austenitic stainless steels. Understanding the microstructural changes ongoing during long time exposure at elevated temperature lays in the basis of the creep resistance properties of stainless steels. The understanding relation between cold work plastic deformation, mechanical properties including creep and the structural changes with the particular attention to precipitated intermetallic phases during long-term high-temperature exposure is essentials to predict the lifetime of engineering applications. The changes of the structure and mechanical properties is extensively studied both for the base austenitic steels grades in annealed conditional, and after cold bending or welding, that can further accelerate precipitation of sigma phase, supplying more deeper view on observed degradation processes [8–10].

The high-temperature service conditions require creep-resistant alloys of the austenitic or ferritic microstructure. The creep-resistant alloys during service undergo progressive degradation of their microstructure, which results in the changes of functional properties. The main microstructural mechanisms of degradation include the processes of matrix softening, the processes of precipitation and matrix depletion of the interstitial and substitution elements. The precipitation processes ongoing during service plays a significant indicator of microstructure degradation processes. Therefore, such processes are essential in diagnostic of

http://dx.doi.org/10.5772/intechopen.72062

components and equipment in service and make possible to forecast the time of safe operation. The characteristics of secondary phase precipitate occurring in creep-resistant steels, especially most frequently used austenitic creep-resistant grades is essential to the understanding of ongoing degradation processes [11, 12].

Apart from high-temperature applications of creep-resistant austenitic alloys, the austenitic stainless steels are frequently applied in low-temperature service conditions. High toughness at the low-temperature of austenitic stainless grades, in contrast to ferritic or duplex alloys, makes them particularly suitable cryogenic applications. The effect of intergranular precipitation on the low-temperature toughness of nitrogen alloyed austenitic stainless steels plays an essential role in cryogenic applications. The presence of intergranular nitride precipitates causes a severe decrease in toughness for stainless steels subjected to sort high-temperature cycle. Nitrogen alloyed stainless steels exhibit the presence of intergranular brittle fracture as a result of the grain boundary nitride precipitation. This behavior may be especially crucial during the welding, because of the short aging time for precipitation. The more profound understanding of precipitation processes involves thermodynamic stability and growth kinetics analysis of the precipitated phases, during a high-temperature cycle. For such a purpose, computational thermodynamics and the so-called CALPHAD method wave been frequently applied [13, 14].

The austenitic stainless steels are not free from corrosion problems. The book also deals with most frequently local corrosion phenomena encountered in these alloys. The local corrosion processes like intergranular corrosion (IGC) occurring in stainless steels remains in the interest of science. Therefore, alloy design strategies also focused on alloying of austenitic stainless steels by single or combined addition of nitrogen, molybdenum, and silicon. Based on such alloying principles, many austenitic stainless steels grades can be defined, dedicated to resisting local corrosion in chloride-containing media and nitric acid [15].

Studies on corrosion resistance also include modification of working conditions, thus the introduction of corrosion inhibitors to an acidic environment. The research on corrosion inhibitors is even going in the direction of organic compounds, for example, rosemary oil and aniline that can significantly reduce the corrosion rate of stainless steels in sulfuric acid [16, 17].

Austenitic stainless steels are well weldable, even though different problems have often been reported during the welding operation. These issues are also addressed in the book. Welding of stainless steel using friction welding becomes more interesting nowadays. The problems encountered in friction welding during joining of austenitic stainless steel are very limited when compared to fusion welding process. Therefore such technic can be useful for joining dissimilar metals, for example, joining of austenitic stainless steel and copper base alloys. Evaluation of friction welding parameter and understanding of ongoing metallurgical phenomenon plays a key role in finding the good bond strength between dissimilar metals [18].

Applications of austenitic stainless steels, apart from corrosion environment action, also involve in mechanical loads. The nature and intensity of such loads can be detrimental for each engineering equipment, thus must be determined and studied through basic microstructural mechanisms. The knowledge of their character is crucial in maintaining the required mechanical properties of the elements and the entire system made of stainless steels. The fracture mechanisms, thus reasons of nucleation and propagation of cracks of different size (from micro- to macro-cracks) under low cycle fatigue conditions is essential to modern high-risk systems, like nuclear applications. For this reason, different grades of austenitic steels as prospective materials for structural parts subjected to extreme cyclic loading and severe environmental conditions have been studied. The nucleation of microcracks in the austenitic, fine-grained steels was observed in slip bands and the critical factor in the destruction of the material is assigned to precipitates and grain boundaries. The microcrack propagation process and its correlation with the steel microstructure are also quite disputable. Despite numerous experimental results on alteration of microcrack propagation direction, there is no convincing hypothesis explaining the real reasons for the process [19, 20].

Author details

Zbigniew Brytan, Wojciech Borek* and Tomasz Tański

*Address all correspondence to: wojciech.borek@polsl.pl

Faculty of Mechanical Engineering, Institute of Engineering Materials and Biomaterials, Silesian University of Technology, Gliwice, Poland

References

[1] Nickel Magazine. Special Issue: Celebrating the 100th Anniversary of Stainless Steel, May 16, 2012

[2] "The Discovery of Stainless Steel." British Stainless Steel Association, Web. 01 Nov. 2017, http://www.bssa.org.uk/about_stainless_steel.php?id=31

[3] The presentation "Stainless Steel and Nickel—Still Partners After 100 Years" given 18 September 2012 by Dr. Peter Cutler, Director Promotion at the 5th International Stainless Steel Symposium in Düsseldorf Germany. Web. 01 Nov. 2017, https://www.nickelinstitute.org/KnowledgeBase/Presentations/StStAndNiPartnersAfter100YearsDusseldorfSept2012.aspx

[4] Mola J, Wendler M, Weiß A, Reichel B, Wolf G, De Cooman BC. Segregation-induced enhancement of low-temperature tensile ductility in a cast high-nitrogen austenitic stainless steel exhibiting deformation-induced α' martensite formation. Metallurgical and Materials Transactions A: Physical Metallurgy and Materials Science. 2015;**46**(4):1450. DOI: 10.1007/s11661-015-2782-y

[5] Mola J, Ullrich C, Kuang B, Rahimi R, Huang Q, Rafaja D, et al. Austenitic nickel- and manganese-free Fe-15Cr-1Mo-0.4N-0.3C steel: Tensile behavior and deformation-induced processes between 298 K and 503 K (25 °C and 230 °C). Metallurgical and Materials

Transactions A: Physical Metallurgy and Materials Science. 2017;**48**:1033-1052. DOI: 10.1007/s11661-017-3960-x

[6] Kaleta J, Tumański S, Żebracki J. Magnetoresistors as a tool for investigation of mechanical properties of ferromagnetic materials. Journal of Magnetism and Magnetic Materials. 1996;**160**:165-166

[7] Jerzy K, Przemysław W. Magnetovisual method for monitoring thermal demagnetization of permanent magnets used in magnetostrictive actuators. Journal of Rare Earths. 2014;**32**(3):236-241

[8] Wilms ME, Gadgil VJ, Krougman JM, Ijsseling FP. The effect of σ-phase precipitation at 800°C on the corrosion resistance in sea-water of a high alloyed duplex stainless steel. Corrosion Science. 1994;**36**(5):871-881

[9] Souza CM, Abreu HFG, SSM T, JMA R. The σ phase formation in annealed UNS S31803 duplex stainless steel: Texture aspects. Materials Characterization. 2008;**59**(9):1301-1306

[10] Kander L, Stejskalová Š, Čížek P. The change of the structure and mechanical properties of the austenitic steels after exposure at the critical temperature. In: Conference Proceedings METAL 2016: 25th Anniversary International Conference on Metallurgy and Materials. ISBN 978-80-87294-67-3. pp. 724-729

[11] Golański G, Lis AK, Słania J, Zieliński A. Microstructural aspect of long term service of the austenitic TP347HFG stainless steel. Archives of Metallurgy and Materials. 2015;**60**:2091-2094

[12] Sourmail T. Precipitation in creep resistant austenitic stainless steel. Materials Science and Technology. 2001;**14**:1-14

[13] Saucedo-Muñoz ML, Liu CS, Komazaki S, Kwon I, Hashida T, Takahashi H, Nakajima H. Evaluation of thermal aging embrittlement of austenitic stainless steels JN1, JJ1 and JK2 by cryogenic small-punch testing. Journal of Materials Research. 2002;**17**:852-858. DOI: http://dx.doi.org/10.1557/JMR.2002.0124

[14] Thermo-Calc Software, Web. 01 Nov. 2017. http://www.thermocalc.com/products-services/software/thermo-calc

[15] Pisarevsky LA, Filippov GA, Lipatov AA. Effect of N, Mo and Si on the resistance to local corrosion of unstabilized Cr-Ni and Cr-Mn-Ni-austenitic steels. Metallurgist. 2016;**60**(7-8):59-66

[16] Loto RT, Oghenerukewe E. Inhibition studies of rosmarinus officinalis on the pitting corrosion resistance 439LL ferritic stainless steel in dilute sulphuric acid. Oriental Journal of Chemistry. 2016;**32**(5):2813-2832. DOI: 10.13005/ojc/320557

[17] Loto RT, Loto RO, Joseph OO, Akinwumi I. Electrochemical studies of the corrosion inhibition property of rosmarinus officinalis on mild steel in dilute sulphuric acid. Journal of Chemical and Pharmaceutical Research. 2015;**7**(7):105-116

[18] Fu L, Du SG. Effects of external electric field on microstructure and property of friction welded joint between copper and stainless steel. Journal of Materials Science. 2006;**41**: 4137-4142

[19] Eterashvili T, Dzigrashvili T. Study of fracture mechanisms at cyclic fatigue of steels used in nuclear reactors I. Steel Research International. 2012;**83**(3):213-217. DOI: 10.1002/srin.201100184

[20] Eterashvili T, Vardosanidze M. A fracture crystallography and anisotropy of propagation of microcracks nucleated in stainless austenitic steels after LCF. Key Engineering Materials. 2006;**324-325**:935-938. DOI: 10.4028/www.scientific.net/KEM.324-325.935

Chapter 2

Considerations in the Design of Formable Austenitic Stainless Steels Based on Deformation-Induced Processes

Javad Mola

Additional information is available at the end of the chapter

http://dx.doi.org/10.5772/intechopen.70939

Abstract

The temperature dependence of tensile elongation in austenitic steels is discussed in view of the relationship between the stacking fault energy and deformation-induced processes. It is shown that the maximum tensile elongation is achieved in the vicinity of $M_d^{\gamma \rightarrow \alpha'}$ temperature. The influence of alloying elements on the temperature dependence of tensile elongation can therefore be analyzed with regard to their influence on the $M_d^{\gamma \rightarrow \alpha'}$ temperature. In this regard, majority of alloying elements including C and N decrease the temperature associated with highest tensile elongation. Due to the high efficiency of C and N in increasing the stability of austenite, approaches toward the development of high-interstitial austenitic stainless steels containing minimal amounts of substitutional alloying elements are discussed. Finally, some of the challenges associated with the processing of high-interstitial austenitic stainless steels are reviewed.

Keywords: deformation-induced processes, transformation-induced plasticity, twinning-induced plasticity, stacking fault energy, tensile elongation, high-interstitial steels

1. Introduction

Although austenitic steels are often inferior to ferritic steels in view of the yield strength (YS), they have the potential to work harden to a larger extent which in turn leads to very high ultimate tensile strength (UTS) values. Due to their low YS/UTS ratios, austenitic steels may be used in applications where a high energy absorption capacity rather than a high resistance to yielding is required. In contrast to ferritic steels of which the strain-hardening behavior is not much influenced by varying temperature in the vicinity of room temperature and by the addition of common substitutional alloying elements such as Mn, Si, Ni, and Cr, the strain-hardening characteristics of austenitic steels are highly sensitive to temperature and chemical composition. Understanding the underlying mechanisms governing the work hardening of austenitic steels and the interrelationships among the material parameters is therefore essential to the effective design of economical austenitic steels. Tremendous austenitic stainless steel design activities at the TU Bergakademie Freiberg, especially in the framework of the Collaborative Research Center 799 (TRIP-matrix composites) running since 2008, have enabled to note the interrelationships summarized in the following sections.

2. Relationship between the stacking fault energy (SFE) and deformation-induced processes

Deformation-induced processes in austenitic steels are governed by the SFE. In high Mn austenitic steels, for instance, deformation processes have been proposed to change from the perfect dislocations glide at high SFEs to twinning and then martensite formation at progressively lower SFEs [1]. The SFE depends on the chemical composition and temperature. Numerous empirical relationships have been proposed to describe the influence of alloying elements in austenitic steels on the SFE [2–6]. Nevertheless, care must be taken when using such relationships because their applicability depends on the reliability of the SFE determination method. The inaccuracy of the existing relationships is reflected in the inconsistent coefficients proposed for alloying elements [2, 6]. Although SFE appears, based on the existing empirical relationships, to have a complex dependence on the chemical composition, it is unanimously known to increase at higher temperatures [7]. The associated changes in the glide mode have important consequences for the deformed microstructures obtained at different temperatures; the wide separation of a/6 <112> Shockley partial dislocations at low temperatures, where the SFE is low, promotes the planar glide of dislocations, for instance by restricting the cross-slip of screw dislocations. The abundance of stacking faults in the microstructure of an Fe-14Cr-5.5Mn-5.5Ni-0.37N (concentrations always in mass-%) austenitic stainless steel deformed at −40°C is demonstrated in **Figure 1**. The occurrence of planar glide features such as dislocation pile-ups, stacking faults, twins, and ε-martensite at deformation temperatures associated with low SFEs highlights the traces of $\{111\}_\gamma$ glide planes. In the example shown in **Figure 2**, the angular relationships among the traces of glide planes aid to determine the crystallographic orientation of austenite.

As the deformation temperature increases, the reduced separation of partial dislocations and the possibility of constriction increase the three-dimensional mobility of dislocations and a gradual transition to the wavy glide mode follows [8]. Deformed microstructures under wavy glide conditions are characterized by dislocation cell structures with almost dislocation-free interiors surrounded by dislocation rotation boundaries [9, 10]. Dislocation cell formation in an Fe-18Cr-7Mn-9Ni-0.43C stainless steel deformed at 200°C is demonstrated in **Figure 3**. Further examples of austenitic stainless steel microstructures formed at various deformation temperatures may be found in [10–16].

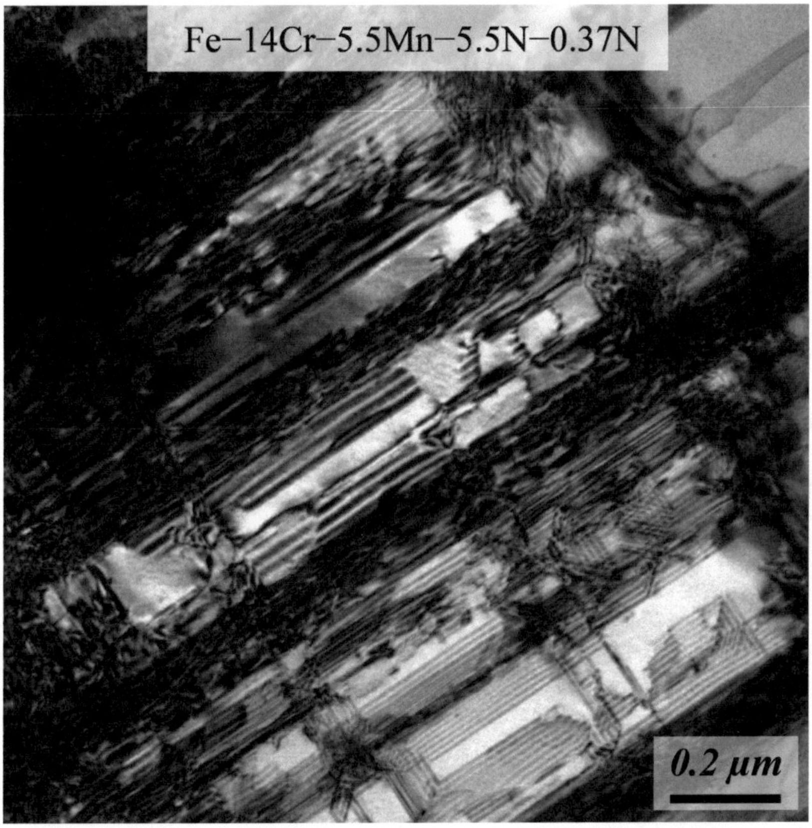

Figure 1. Stacking faults formed in an Fe-14Cr-5.5Mn-5.5Ni-0.37N steel after 64% tensile deformation at −40°C.

The schematic in **Figure 4** summarizes the temperature dependence of deformation-induced microstructural changes in austenitic steels. The spectrum of processes shown in the schematic does not necessarily occur in all austenitic steels. Whether or not a given process is activated is decided by the alloy system. Highly stable austenitic steels may not exhibit the microstructural changes relying on very low SFEs. For example, the austenitic stainless steel Fe-18Cr-20Ni with a high stability does not experience deformation-induced α'-martensite formation even after the application of almost 100% tensile elongation at −196°C [14, 17]. Furthermore, the tendency of an austenitic steel to form ε-martensite or twinning depends on the alloy chemistry. In high Mn steels, for instance, alloying with Si increases the likelihood of ε-martensite formation at the expense of deformation twinning [18]. **Table 1** summarizes the types of deformation-induced microstructural changes (twinning and ε-martensite) in modifications of FeCrMnNi-based austenitic stainless steels. Some of the alloys listed in **Table 1** are twinning only or ε-martensite only systems. Nevertheless, a majority of austenitic stainless steels are capable of exhibiting both the preceding deformation processes as byproducts of planar glide, caused by the ready dissociation of Shockley partial dislocations [19, 20].

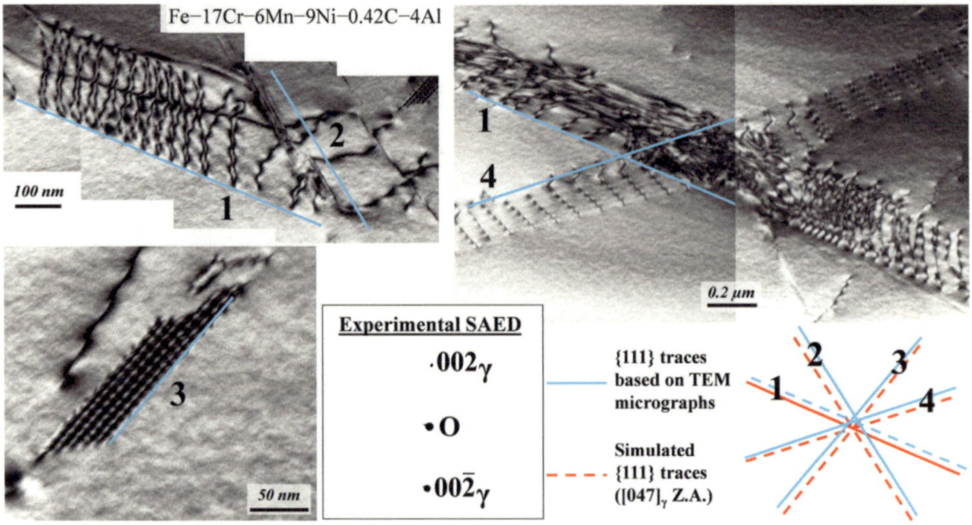

Figure 2. TEM micrographs and the corresponding selected-area electron diffraction (SAED) pattern of an Al-alloyed Fe-17Cr-6Mn-9Ni-0.42C-4Al steel after the application of nearly 4% compressive strain at RT. The zone axis (Z.A.) of austenite cannot be determined from the recorded SAED pattern which only comprises two 002 diffraction spots. Instead, the angular relationships among the traces of $\{111\}_\gamma$ glide planes (superimposed solid lines), which may be identified by the glide plane features, allow to determine the zone axis (Z.A. // $<047>_\gamma$).

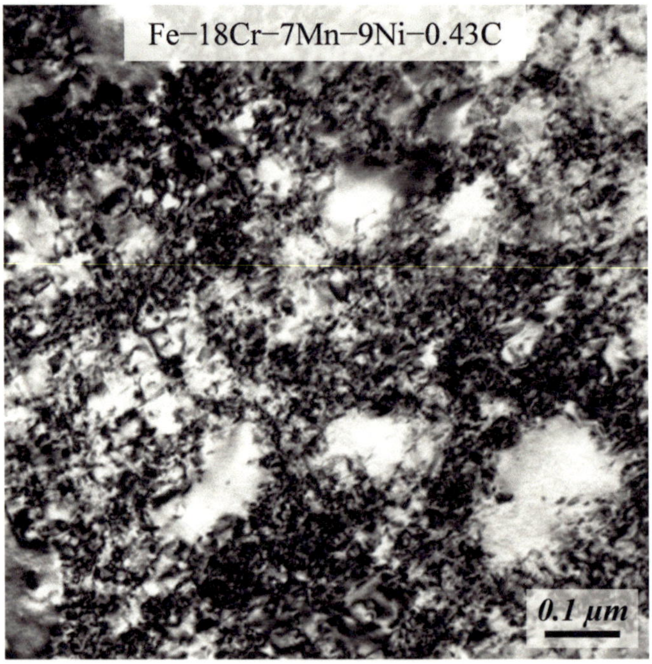

Figure 3. Microstructure of an Fe-18Cr-7Mn-9Ni-0.43C steel after 45% tensile deformation at 200°C.

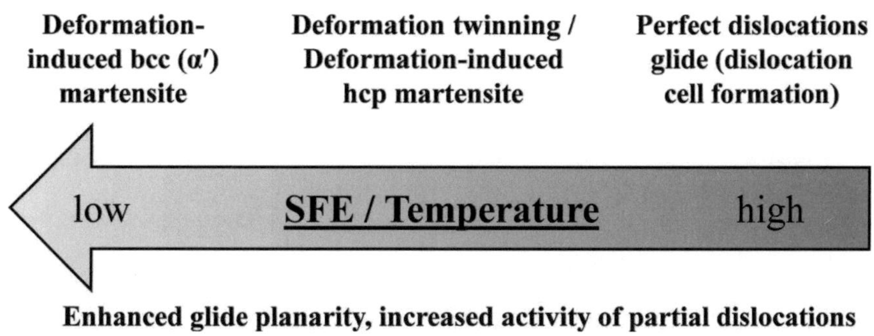

Figure 4. Schematic representation of the effects of SFE and temperature on the microstructural changes in austenitic steels.

Alloy	Twinning	ε-Martensite
Fe-16Cr-(6–7)Mn-(6–9)Ni [21, 22]	✓	✓
Fe-(17–18)Cr-(6–7)Mn-9Ni-0.4C-(0 & 4)Al [15]	✓	–
Fe-14Cr-5.5Mn-5.5Ni-0.37N [14]	–	✓
Fe-15Cr-3Mn-3Ni-(0.05–0.25)C-(0.1–0.13)N [23, 24]	✓	✓
Fe-19Cr-3Mn-4Ni-(0.05–0.25)C-(0.1–0.13)N [25–27]	✓	✓
Fe-15Cr-1Mo-0.3C-0.4N [28]	✓	–

Table 1. Occurrence of twinning/ε-martensite formation as a function of chemical composition (mass-%).

Deformation at temperatures lower than those associated with ε-martensite formation and twinning leads to the formation of α'-martensite. As mentioned earlier, this does not apply to highly stable austenitic steels in which the threshold SFEs for the α'-martensite formation may not be attained even at cryogenic temperatures [14]. The α'-martensite formation under extremely planar glide conditions commonly occurs at the intersections of planar glide features [29–31]. Examples of deformation-induced α'-martensite formation in conjunction with planar glide features are shown in **Figure 5**. Even in cases where it is difficult to identify intersection of planar glide features, such as the example of **Figure 5d** where the α'-martensite appears to have formed in association with deformation twins, the activation of at least one conjugate partial dislocation glide system is quite likely. After all, the most widely accepted models proposed for the α'-martensite formation rely on the occurrence of two shears in directions compatible with the glide direction of partial dislocations [32–35]. The magnitudes of the shears needed to cause the lattice change to bcc/bct are only fractions of the full twinning shear. Even the formation of spontaneous α'-martensite appears to be a consequence of stacking fault interactions. As the microstructure of the Fe-15Cr-1Mo-0.3C-0.4N steel after partial transformation to athermal α'-martensite shows (**Figure 6**), a high density of stacking faults exists in the austenite, in particular in the immediate vicinity of the transformed regions. Therefore, further transformation of austenite most likely involves the spontaneous dissociation and interaction of the stacking faults. This suggests that the SFE must be extremely low at the α'-martensite start (M_s) temperature.

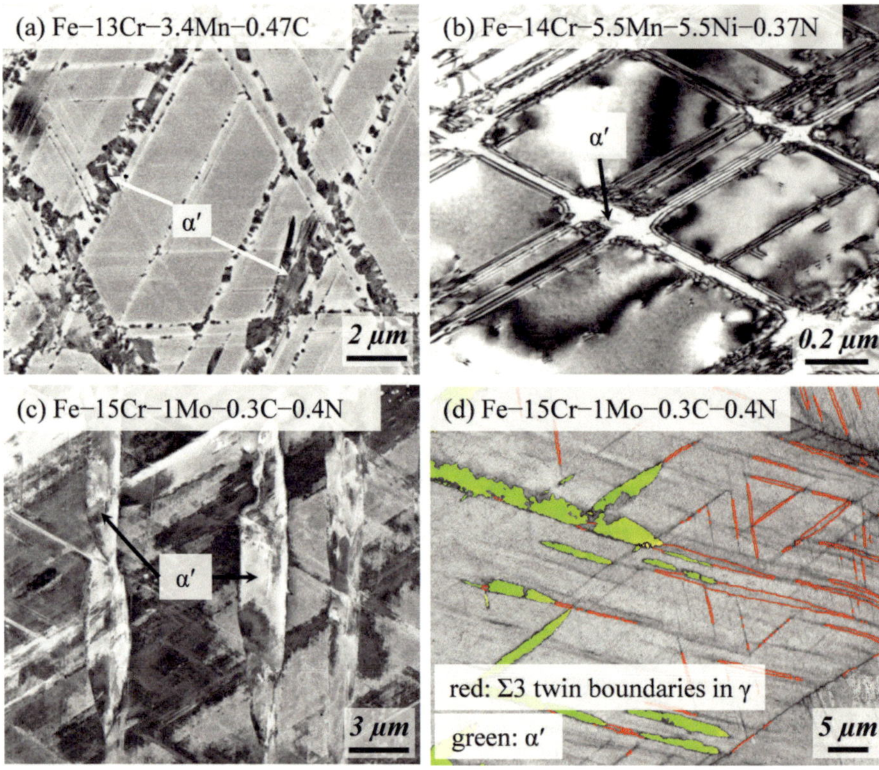

Figure 5. Examples of deformation-induced α'-martensite formation in high carbon (a), high nitrogen (b), and high carbon and nitrogen (d) austenitic stainless steels.

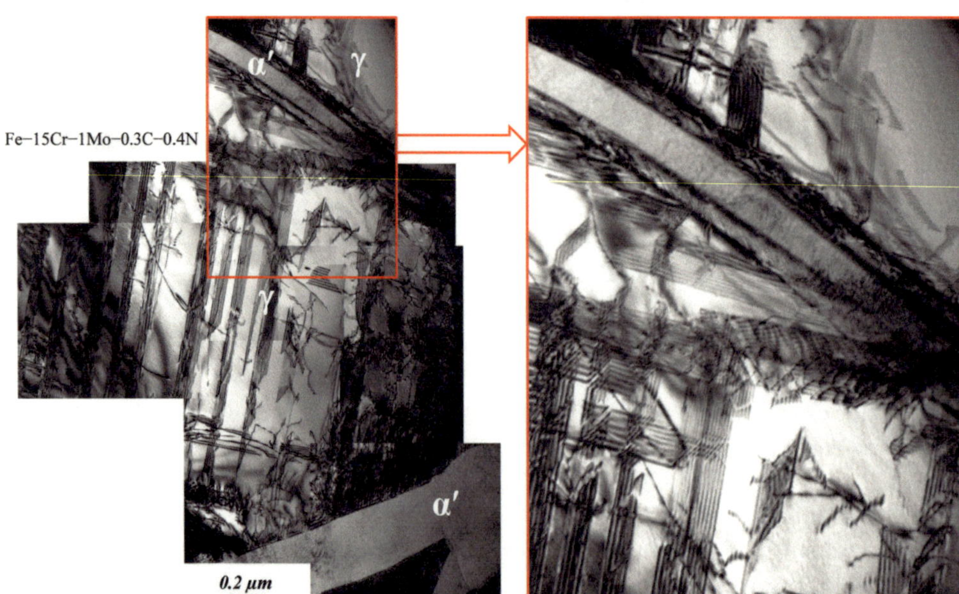

Figure 6. Microstructure of an Fe-15Cr-1Mo-0.3C-0.4N steel after partial transformation to spontaneous (athermal) α'-martensite by thermal treatment.

3. Influence of deformation-induced processes on the strain-hardening behavior and tensile elongation

An important consequence of decreasing the tensile temperature of austenitic stainless steels is the increase in the strain-hardening rate. This is demonstrated using the true stress-strain and strain-hardening curves for an Fe-18Cr-20Ni steel (**Figure 7**) [17]. The enhancement of strain-hardening rate and its extended near-linear hardening at low temperatures can be explained by the decrease in the SFE and the increased activity of stacking faults. At −196°C, the strain-hardening rate remains almost constant until fracture. Near-linear strain hardening during tensile deformation is the characteristic of many austenitic stainless steels at temperatures where SFE is sufficiently low to ensure a high glide planarity but not low enough to enable the formation of α'-martensite. **Figure 8** shows the true strain-hardening curves for an Fe-16Cr-8Mn-7Ni-1Si steel deformed at various temperatures [36]. At temperatures below $M_d^{\gamma \rightarrow \alpha'}$ temperature which is between 20 and 60°C, the increase in the work-hardening rate characteristic of deformation-induced α'-martensite formation can be readily identified [37]. The strain-hardening curves indicate that near-linear strain hardening in austenitic steels is achieved at temperatures near $M_d^{\gamma \rightarrow \alpha'}$ temperature [15].

The schematic in **Figure 9** generalizes the temperature dependence of tensile elongation in austenitic stainless steels [10, 15, 38–43]. The schematic elongation curve consists of three regions marked I–III. In region I, the elongation shows a weak temperature dependence which resembles that of materials with wavy glide, for example, ferritic steels. The region marked II is characterized by enhanced low-temperature ductility due to the enhanced glide planarity. The enhancement of ductility by lowering temperature in region II is usually attributed to deformation-induced microstructural changes such as ε-martensite transformation, the ε-TRIP effect [44], and deformation twinning, the TWIP effect [38]. The hcp stacking of close-packed planes in ε-martensite may be achieved by the glide of 1/6 <112> Shockley partials on every second {111} plane [45]. Deformation twins, on the other hand, can occur by the glide of 1/6 <112> Shockley partials on successive {111} planes [46, 47]. The region marked III in **Figure 9** can appear at deformation temperatures below $M_d^{\gamma \rightarrow \alpha'}$ temperature. The temperature-dependent true stress-strain curves in **Figures 10 (a,b)** and the associated α'-martensite fractions at the end of tensile tests

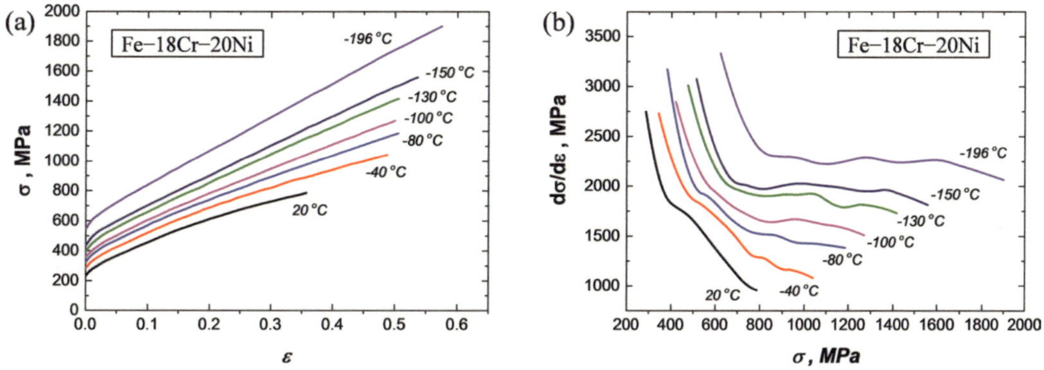

Figure 7. (a) Temperature dependence of true stress-strain curves for an Fe-18Cr-20Ni steel; (b) true strain-hardening curves corresponding to (a) (adapted from Ref. [17]).

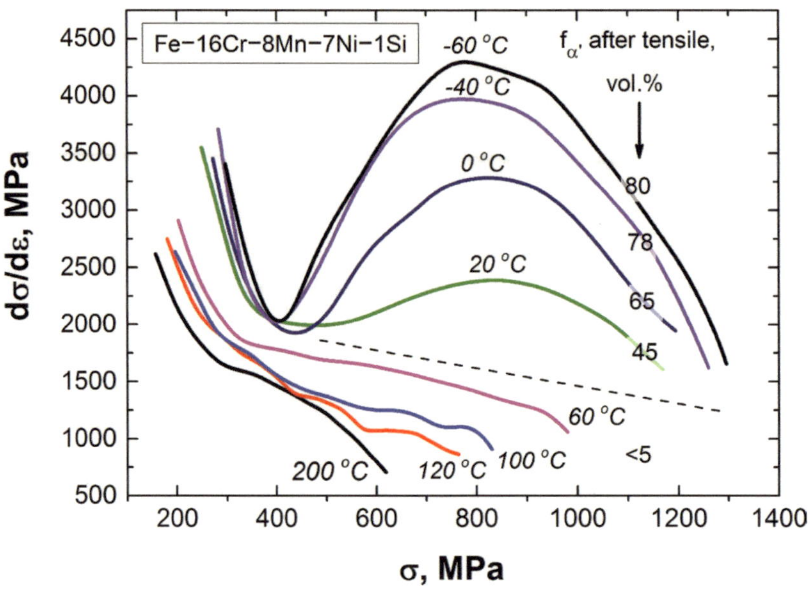

Figure 8. Temperature dependence of true strain-hardening curves for an Fe-16Cr-8Mn-7Ni-1Si steel. The martensite fractions after tensile tests are also marked (adapted from Ref. [36]).

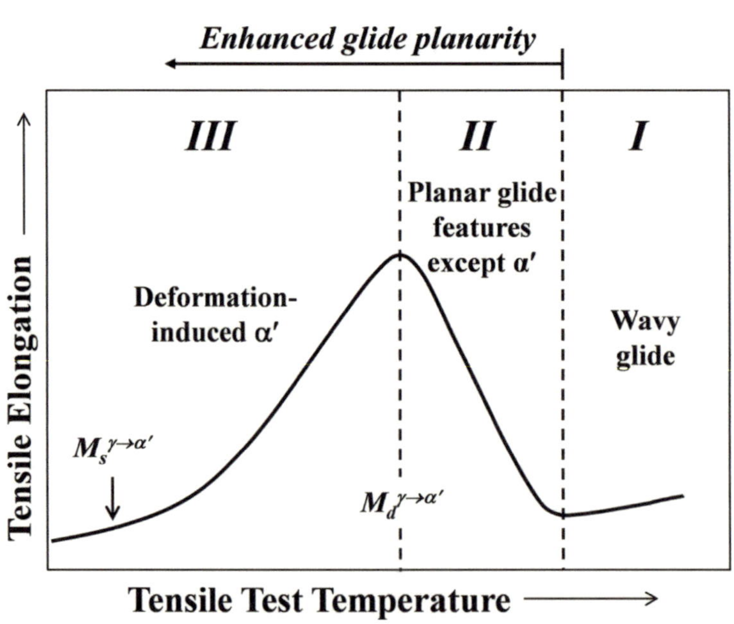

Figure 9. Generalized temperature dependence of tensile elongation in austenitic steels.

[36] (**Figure 10c**) confirm the negative impact of deformation-induced α'-martensite formation on the tensile ductility. Accordingly, the region III of elongation curves is absent in highly stable austenitic steels which do not undergo deformation-induced α'-martensite formation (e.g., the Fe-18Cr-20Ni alloy in **Figure 7**). The loss of ductility in region III in spite of the enhanced

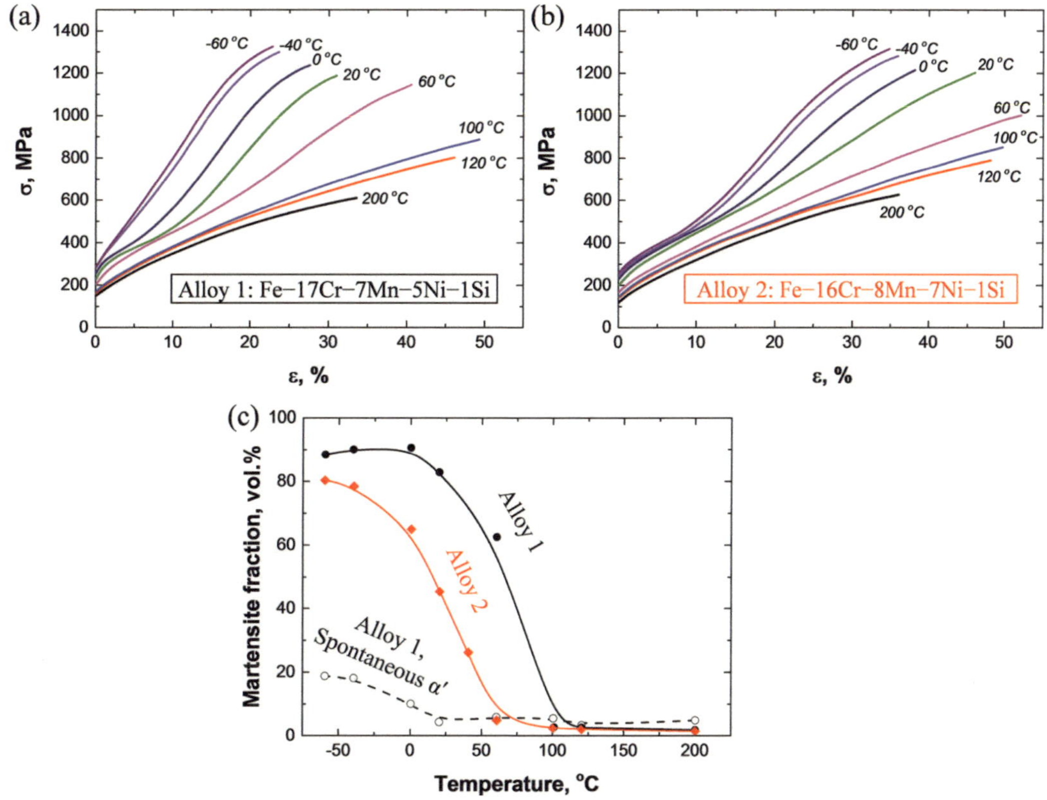

Figure 10. (a and b) Temperature dependence of true stress-strain curves for the marked steel compositions; (c) deformation-induced martensite fractions after tensile tests. The spontaneous martensite fractions are also given for alloy 1 (adapted from Ref. [36]).

glide planarity of austenite is due to the gradual replacement of austenite by the less ductile α'-martensite.

Although the proximity of $M_d^{\gamma\rightarrow\alpha'}$ temperature and the temperature associated with maximum tensile elongation has been documented in many researches involving tensile tests at various temperatures followed by the quantification of bulk α'-martensite fractions in the uniformly strained regions of tensile specimens [14, 17, 36–39], there also exist experimental results in support of the ductility-enhancing effect of α'-martensite formation [10, 14]. **Figure 11** shows such an instance where the tensile elongation increases concurrently with the formation of α' at temperatures below $M_d^{\gamma\rightarrow\alpha'}$ temperature. This observation has been justified by the non-uniform distribution of alloying elements, originating from the solidification step [14]. During solidification, alloying elements tend to segregate from the dendrite cores into the surrounding liquid, namely interdendritic regions in the final microstructure [48] (**Figure 12**). The severity of segregation depends mainly on the cooling rate during solidification. Nevertheless, the solidification mode (austenitic/ferritic) should also be relevant to the extent of solidification segregation. The compositional difference between the dendritic and interdendritic regions could lead to the inhomogeneity of material parameters such as SFE and $M_d^{\gamma\rightarrow\alpha'}$ temperature. Upon plastic deformation, the dendritic and interdendritic regions will exhibit deformation-induced processes commensurate with

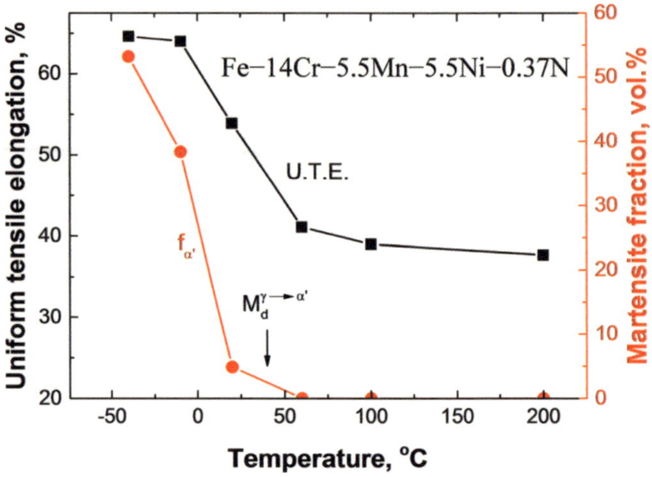

Figure 11. Temperature dependence of uniform tensile elongation for an Fe-14Cr-5.5Mn-5.5Ni-0.37N steel and the associated deformation-induced martensite fractions (adapted from Ref. [14]).

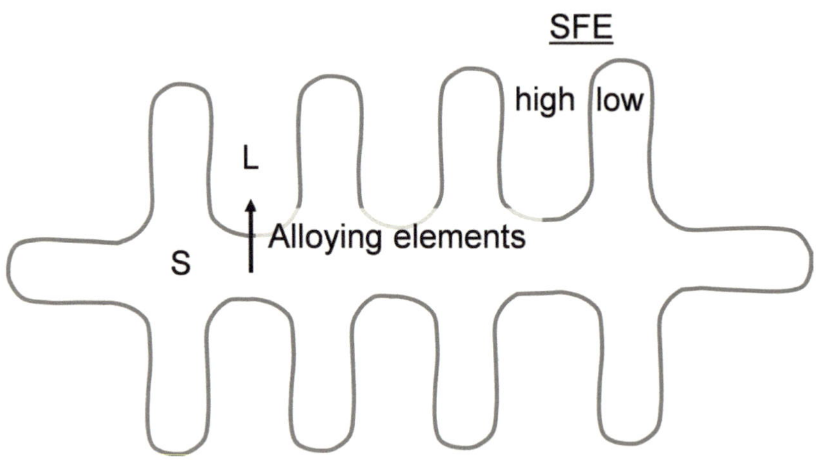

Figure 12. Schematic representation of the segregation of alloying element during solidification. The solid (S) and liquid (L) regions will appear as dendritic and interdendritic regions with different SFE values in the final microstructure.

their SFE [49]. For instance, the dendritic regions with a leaner chemical composition and a lower SFE compared to the interdendritic regions will exhibit a higher glide planarity. In other words, as the deformation temperature decreases, the formation of stacking fault, twins/ε-martensite, and α' is first triggered in the dendritic regions [14]. The presence of a gradient of chemical composition in the microstructure might be responsible for the reported plasticity-enhancing effect of α'-martensite formation, the so-called TRIP effect. To visualize how the inhomogeneity of chemical composition can mask the negative impact of deformation-induced α'-martensite formation on the tensile elongation, the temperature dependence of tensile elongation and deformation-induced α'-martensite fractions for two steels with slight differences in the chemical composition are summarized in **Figure 13** [36]. The steels Fe-16Cr-7Mn-6Ni-1Si and Fe-16Cr-8Mn-7Ni-1Si can

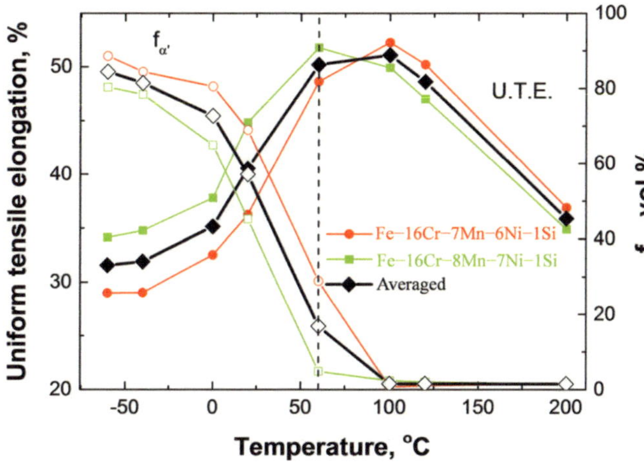

Figure 13. Temperature dependence of uniform tensile elongation for two steels with slightly different chemical compositions and the averaged values (adapted from Ref. [36]).

represent the chemical compositions of dendritic and interdendritic regions of segregated steel, respectively. The tensile elongation of a hypothetical alloy consisting of equal fractions of these two steels is estimated in **Figure 13** by averaging (rule of mixtures). The rule of mixtures was also applied to determine the temperature dependence of deformation-induced α'-martensite fraction in the hypothetical alloy. As the dashed line in **Figure 13** shows, the tensile elongation of the segregated hypothetical alloy is not much impaired by the formation of approximately 17 vol.% deformation-induced α'-martensite. In the presence of still larger gradients of chemical composition, the tensile ductility might even increase in spite of the formation of α'-martensite in the least stable regions of the microstructure. This can indeed take place if the negative impact of α'-martensite formation in the less stable regions of the microstructure is more than neutralized by the positive effect of planar glide enhancement in the more stable regions of the microstructure [14]. Based on the preceding discussion, the recognition of the α'-martensite formation as a plasticity-enhancing mechanism (α'-TRIP effect) in TRIP-assisted [50] and quenched and partitioned (Q&P) [51] steels containing small quantities of austenite is possibly due to local variations in the chemical composition of austenite.

4. Design of austenitic stainless steels

As shown in the schematic of **Figure 9**, highest tensile elongations for austenitic steels with a uniform distribution of alloying elements may be obtained over a narrow temperature range near $M_d^{\gamma\rightarrow\alpha'}$ temperature. The objective of many austenitic steel developments is therefore to ensure that the deformation temperature or the service temperature, where a high formability or a high-energy absorption capacity is required, lies in the vicinity of the peak elongation temperature. It is because of formability considerations that the $M_d^{\gamma\rightarrow\alpha'}$ temperature of commercial austenitic stainless steels is often in the vicinity of room temperature [52, 53]. Therefore, the $M_d^{\gamma\rightarrow\alpha'}$ temperature may be used to predict the influence of alloying elements on the elongation curve. Tensile tests at various temperatures of austenitic stainless steels with varied amounts of Ni [39, 42], Mn [43], Al [15], Cr [23, 25, 27], C [23, 24], and N [25] have confirmed that they shift the peak elongation temperature to lower temperatures. All of the preceding alloying elements except Al are unanimously known to decrease both $M_d^{\gamma \rightarrow \alpha'}$ and M_s temperatures [54, 55]. The formation of AlN and the associated reduction in the solute N content might be responsible for the reported increase in the M_s temperature upon the addition of only small quantities of Al. This effect is expected to disappear when Al is present in quantities much higher than the stoichiometric Al content needed for the full stabilization of solute N. For instance, the addition of 4.9 mass-%Al to a cast iron has decreased the M_s temperature [56].

Economical design of highly formable austenitic stainless steels, namely lean austenitic stainless steel compositions, should be based on the principle that all of the elements which decrease the $M_d^{\gamma \rightarrow \alpha'}$ temperature, irrespective of their reported influence on the SFE, also decrease the peak elongation temperature. This is shown schematically in **Figure 14**. To secure a high formability at room temperature, the $M_d^{\gamma \rightarrow \alpha'}$ temperature must be close to room temperature as is the case with commercial austenitic stainless steels such as the AISI 304-grade (Fe-18Cr-10Ni) stainless steel. The development of the AISI 200 series (FeCrNiMn) austenitic stainless steels in which Ni is partially replaced with Mn is also in accord with the preceding design principle [39].

Since C and N effectively decrease the $M_d^{\gamma \rightarrow \alpha'}$ temperature, an economical design approach would be to partially replace the substitutional alloying elements Cr, Ni, and Mn with the interstitials C and N. The maximum level of N, which can be introduced into the molten steel, depends on both the chemistry and the N_2 partial pressure applied during casting and melting [57]. Stainless steel compositions Fe-15Cr-3Mn-3Ni-0.1N–(0.05–0.25)C exemplify the efforts made to reduce the substitutionals content by the addition of both C and N [23, 24]. Nevertheless, even for the alloy containing the highest C content of 0.25 mass-%, the stability

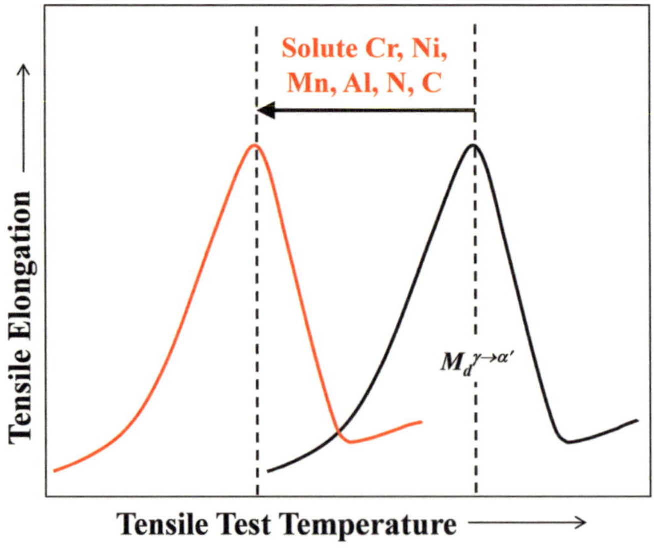

Figure 14. Influence of alloying elements on the temperature dependence of tensile elongation.

of austenite at room temperature was not sufficiently high to suppress the early deformation-induced occurrence of α'-martensite. As a result, the room temperature tensile elongation remained below 15%. At 200°C, where the deformation-induced α'-martensite formation was inhibited, the total elongation increased to about 65%. In comparison, the Fe-19Cr-4Ni-3Mn-0.15N–(0.05–0.25)C steels with higher Cr, Ni, and N contents exhibited tensile elongations up to 50% at room temperature [25].

As long as the full dissolution of carbides and nitrides is enabled, still higher quantities of interstitials may be added to stainless steels to reduce the required amounts of substitutional

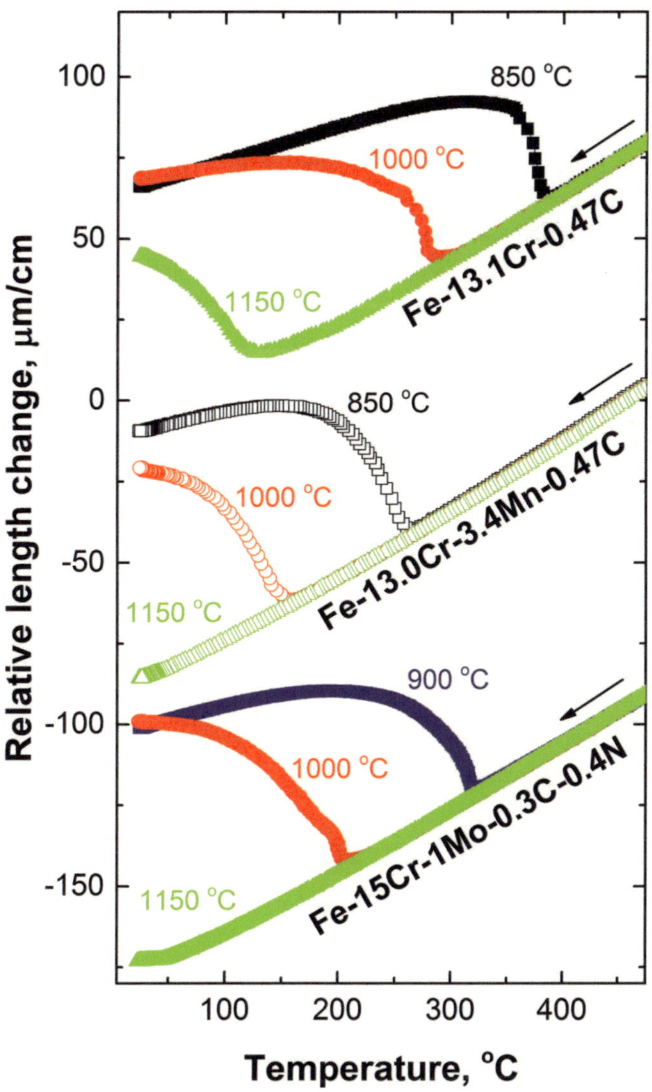

Figure 15. Relative length changes during cooling of Fe-13.1Cr-0.47C, Fe-13.0Cr-3.4Mn-0.47C, and Fe-15Cr-1Mo-0.3C-0.4N steels from the indicated solution-annealing temperatures. As the solution-annealing temperature increases, the M_s temperature shifts to lower temperatures.

alloying elements. Interstitial alloying elements are also effective solid solution strengtheners of austenite. Fe-18Cr-10Mn-base steels alloyed with up to 0.69 mass-%N are examples of N-alloyed stainless steels free of Cr_2N precipitates at the solution-annealing temperature [58, 59]. Alloying with higher amounts of N or very high levels of (C + N) will increase the solution-annealing temperature needed for the full dissolution of carbides and nitrides. On this basis, the complete dissolution of carbides and nitrides in the stainless steel Fe-19Cr-19Mn-0.49C-0.58N [60] appears improbable. In fact, the presence of Cr-rich carbides and nitrides is not desirable since Cr, C, and N will then not contribute to the stabilization of austenite. Furthermore, the corrosion resistance is impaired by the presence of Cr-rich precipitates. Assuming that all interstitials in an alloy such as the Fe-19Cr-19Mn-0.49C-0.58N steel are taken into solution by a high-temperature solution-annealing treatment, the $M_d^{\gamma \rightarrow \alpha'}$ temperature and the peak elongation temperature will be far lower than room temperature.

An alternative approach to the design of lean-alloy austenitic stainless steels is to modify the chemical composition and/or the thermal processing of standard quenched and tempered martensitic stainless steels to make them fully austenitic at room temperature. The room temperature microstructure of an Fe-13.1Cr-0.47C steel after solution annealing at 1250°C consisted of 44 vol.% α'-martensite and 56 vol.% austenite. As shown in the dilatometry curves of **Figure 15**, the addition of 3.4 mass-%Mn to the alloy decreased the M_s temperature to around room temperature such that a fully austenitic microstructure was obtained. $M_d^{\gamma \rightarrow \alpha'}$ temperature for the latter Mn-added steel is much higher than room temperature. Accordingly, significant enhancement of tensile ductility was observed at 200°C (**Figure 16**). A similar scenario arises in the case of steels containing both C and N, for instance the commercial Fe-15Cr-1Mo-0.4N-0.3C steel which is conventionally used in the martensitic condition [28]. After solution annealing at 1250°C which resulted in a fully austenitic microstructure, ten-

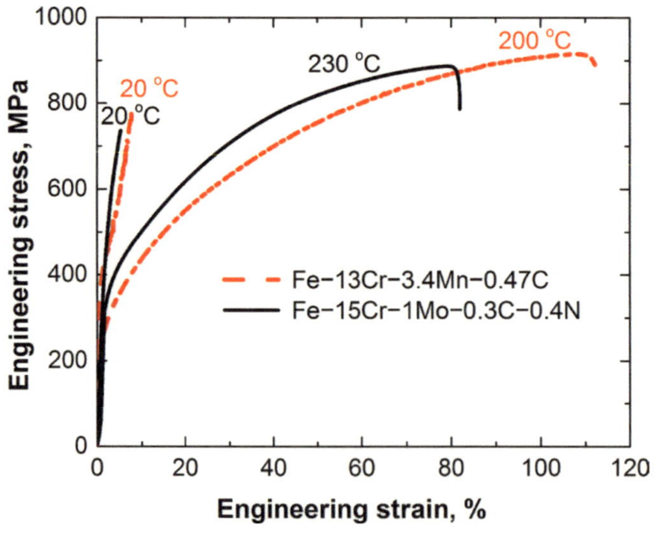

Figure 16. Engineering stress-strain curves for Fe-13Cr-3.4Mn-0.47C (dashed lines) and Fe-15Cr-1Mo-0.4N-0.3C (solid lines) steels at the indicated temperatures (adapted from [28, 61]).

sile elongation at room temperature was less than 5%. Tensile elongation was increased to 80% at 230°C (**Figure 16**). Such lean austenitic stainless steel compositions may be used for warm-forming operations at temperatures around 200°C. In the case of both Fe-13Cr-3.4Mn-0.47C and Fe-15Cr-1Mo-0.4N-0.3C stainless steels, decreasing the $M_d^{\gamma \rightarrow \alpha'}$ temperature to room temperature by raising the substitutionals content would be a possible way to enhance the room temperature tensile ductility.

5. Challenges associated with the processing of high-interstitial steels

The micrographs in **Figure 17** show that the complete dissolution of high levels of C and N such as those contained in the Fe-15Cr-1Mo-0.4N-0.3C alloy requires the use of high solution-annealing temperatures. In the presence of precipitates, a higher amount of substitutional alloying elements will be required to ensure a sufficiently low $M_d^{\gamma \rightarrow \alpha'}$ temperature. Furthermore,

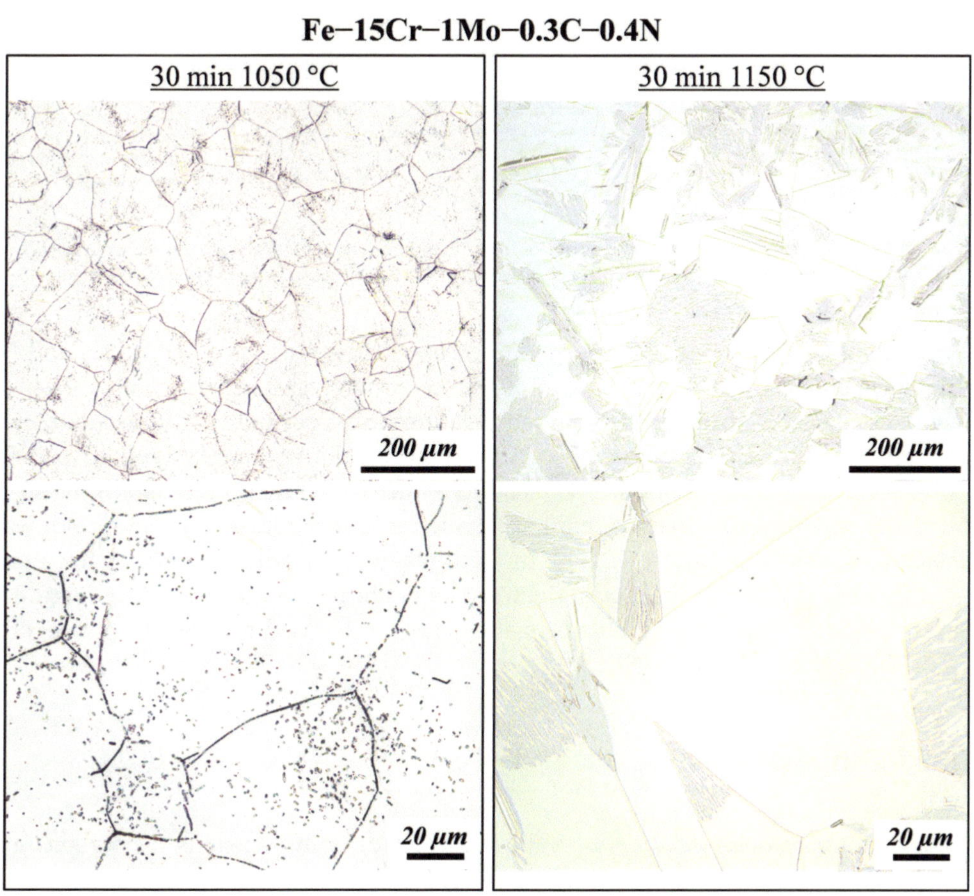

Figure 17. Optical micrographs of Fe-15Cr-1Mo-0.4N-0.3C steel after the indicated solution-annealing treatments. Oxalic acid was used as etchant.

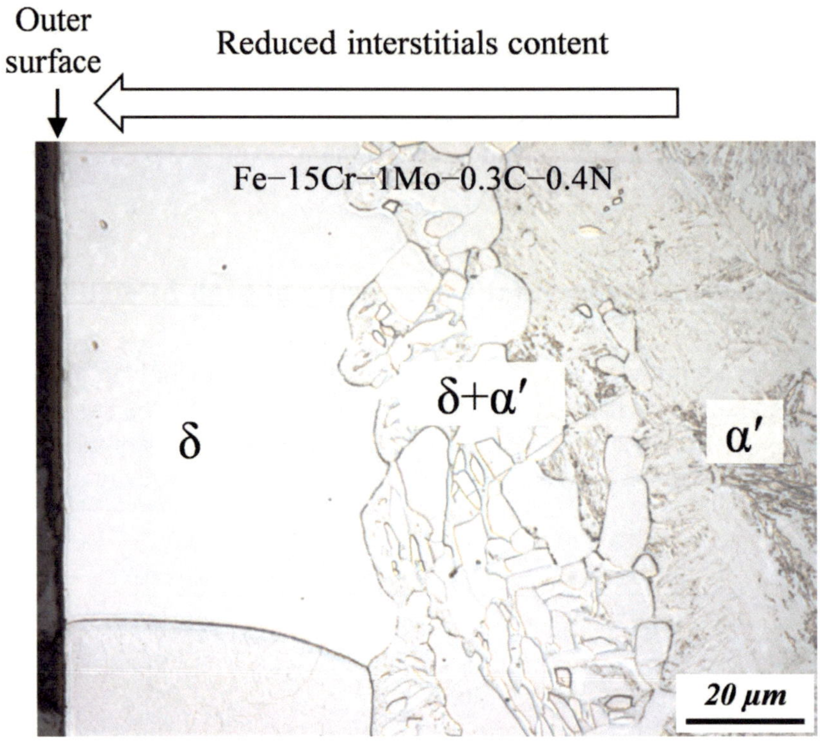

Figure 18. Optical micrograph of a section near the exposed surface of a dilatometry specimen of Fe-15Cr-1Mo-0.4N-0.3C steel after 13 heat treatment cycles at temperatures between 900 and 1180°C. V2A reagent was used as etchant.

the presence of Cr-rich carbides, nitrides, and carbo-nitrides such as $M_{23}C_6$, M_7C_3, and M_2N (M denotes mainly Cr and Fe) will deteriorate the corrosion resistance of stainless steels. The inevitable use of high solution-annealing temperatures increases the risk of decarburization, denitriding, and other surface problems such as uniform and intergranular types of oxidation. The loss of C and N near the surface will lead to a microstructure gradient from the surface toward inner regions. Commonly, a martensitic microstructure forms near the surface where the interstitials content is lowest [23, 28]. In cases where the ferrite potential of the alloy in the absence of interstitials is high, even delta ferrite might form near the surface (**Figure 18**). Surface-to-bulk micro hardness profile measurements can be used to determine the depth of decarburization/denitriding [23, 28].

6. Concluding remarks

The occurrence of various deformation-induced processes such as perfect dislocations glide, deformation twinning, and deformation-induced ε/α′-martensite formation mechanisms in austenitic steels is governed by the SFE. As long as the SFE is known, it should be possible to decide which deformation-induced mechanism is activated during deformation. The SFE determination by experimental diffraction-based techniques (including TEM) is a time-consuming task and requires careful specimen preparation. Therefore, it would be useful to correlate SFE with material parameters which can be determined more readily and are less sensitive to artifacts arising from the specimen preparation. One such material parameter is the $M_d^{\gamma\rightarrow\alpha'}$ temperature which can be determined by tensile tests at various temperatures followed by the verification of α'-martensite formation for instance by magnetic measurements. The likelihood of a correlation between SFE and $M_d^{\gamma\rightarrow\alpha'}$ temperature can be visualized by recalling that the formation of α'-martensite is only enabled if the SFE is sufficiently low to enable the deformation-induced reaction of partial dislocations on intersecting $\{111\}_\gamma$ planes which is a requirement for the fcc → bcc transformation. Therefore, the influence of alloying elements on the SFE of austenitic stainless steels can be deduced from their influence on the $M_d^{\gamma\rightarrow\alpha'}$ temperature. In contrast to the coefficients of alloying elements in the empirical equations proposed for the SFE estimation which are subject to a large scatter, the coefficients of alloying elements in the empirical relationships for the estimation of $M_d^{\gamma\rightarrow\alpha'}$ temperature are fairly consistent. Therefore, relationships giving the compositional dependence of $M_d^{\gamma\rightarrow\alpha'}$ temperature can be used as guidelines for the economical design of austenitic stainless steels. Due to the high efficiency of C and N in increasing the stability of austenite, they can be used to obtain austenite with an appropriate stability. The high temperatures required for the dissolution of carbides, nitrides, and carbo-nitrides and the tendency of C and N to escape from free surfaces are some of the challenges associated with the processing of high-interstitial stainless steels compared to that of conventional stainless steels.

Acknowledgements

The financial support of the German Research Foundation (Deutsche Forschungsgemeinschaft, DFG) in the framework of the Collaborative Research Center 799 (CRC799) and the research grants MO 2580/1-1 and MO 2580/2-1 is gratefully acknowledged. Thanks are also due to the scientific and technical staff involved in the above projects.

Author details

Javad Mola

Address all correspondence to: javad.mola@iest.tu-freiberg.de

Institute of Iron and Steel Technology, Technische Universität Bergakademie Freiberg, Freiberg, Germany

References

[1] Allain S, Chateau J-P, Bouaziz O, Migot S, Guelton N. Correlations between the calculated stacking fault energy and the plasticity mechanisms in Fe–Mn–C alloys. Materials Science and Engineering A. 2004;**387-389**:158-162. DOI: 10.1016/j.msea.2004.01.059

[2] Schramm RE, Reed RP. Stacking fault energies of seven commercial austenitic stainless steels. Metallurgical Transactions A. 1975;**6**:1345-1351. DOI: 10.1007/BF02641927

[3] Rhodes CG, Thompson AW. The composition dependence of stacking fault energy in austenitic stainless steels. Metallurgical Transactions A. 1977;**8**:1901-1906. DOI: 10.1007/BF02646563

[4] Lo KH, Shek CH, Lai JKL. Recent developments in stainless steels. Materials Science & Engineering R: Reports. 2009;**65**:39-104. DOI: 10.1016/j.mser.2009.03.001

[5] Yonezawa T, Suzuki K, Ooki S, Hashimoto A. The effect of chemical composition and heat treatment conditions on stacking fault energy for Fe-Cr-Ni austenitic stainless steel. Metallurgical and Materials Transactions A: Physical Metallurgy and Materials Science. 2013;**44**:5884-5896. DOI: 10.1007/s11661-013-1943-0

[6] Brofman PJ, Ansell GS. On the effect of carbon on the stacking fault energy of austenitic stainless steels. Metallurgical Transactions A. 1978;**9**:879-880. DOI: 10.1007/BF02649799

[7] Rémy L, Pineau A, Thomas B. Temperature dependence of stacking fault energy in close-packed metals and alloys. Materials Science and Engineering. 1978;**36**:47-63. DOI: 10.1016/0025-5416(78)90194-5

[8] Kestenbach H-J. The effect of applied stress on partial dislocation separation and dislocation substructure in austenitic stainless steel. Philosophical Magazine. 1977;**36**:1509-1515. DOI: 10.1080/14786437708238531

[9] Roychowdhury S, Kain V, Neogy S, Srivastava D, Dey GK, Prasad RC. Understanding the effect of nitrogen in austenitic stainless steel on the intergranular stress corrosion crack growth rate in high temperature pure water. Acta Materialia. 2012;**60**:610-621. DOI: 10.1016/j.actamat.2011.09.053

[10] Pozuelo M, Wittig JE, Jiménez JA, Frommeyer G. Enhanced mechanical properties of a novel high-nitrogen Cr-Mn-Ni-Si austenitic stainless steel via TWIP/TRIP effects. Metallurgical and Materials Transactions A: Physical Metallurgy and Materials Science. 2009;**40**:1826-1834

[11] Breedis JF. Influence of dislocation substructure on the martensitic transformation in stainless steel. Acta Metallurgica. 1965;**13**:239-250. DOI: 10.1016/0001-6160(65)90201-4

[12] Abrassart F. Stress-induced $\gamma \rightarrow \alpha$ martensitic transformation in two carbon stainless steels. Application to trip steels. Metallurgical and Materials Transactions A. 1973;**4**:2205-2216. DOI: 10.1007/BF02643289

[13] Vogt J-B, Magnin T, Foct J. Effective stresses and microstructure in cyclically deformed 316l austenitic stainless steel: Effect of temperature and nitrogen content. Fatigue and Fracture of Engineering Materials and Structures. 1993;**16**:555-564. DOI: 10.1111/j.1460-2695.1993.tb00766.x

[14] Mola J, Wendler M, Weiß A, Reichel B, Wolf G, De Cooman BC. Segregation-induced enhancement of low-temperature tensile ductility in a cast high-nitrogen austenitic

stainless steel exhibiting deformation-induced α' martensite formation. Metallurgical and Materials Transactions A: Physical Metallurgy and Materials Science. 2015;**46**(4): 1450. DOI: 10.1007/s11661-015-2782-y

[15] Rahimi R, Ullrich C, Klemm V, Rafaja D, De Cooman BC, Biermann H, et al. Influence of Al on the temperature dependence of strain hardening behavior and glide planarity in Fe–Cr–Ni–Mn–C austenitic stainless steels. Materials Science and Engineering A. 2016;**649**:301-312. DOI: 10.1016/j.msea.2015.10.005

[16] Rahimi R, Ullrich C, Rafaja D, Biermann H, Mola J. Microstructural evolution of an Al-alloyed duplex stainless steel during tensile deformation between 77 K and 473 K (−196 °C and 200 °C). Metallurgical and Materials Transactions A: Physical Metallurgy and Materials Science. 2016;**47**:2705-2716. DOI: 10.1007/s11661-016-3438-2

[17] Gutte H, Weiß A. Spannungs- und verformungsinduzierte Martensitbildungen in metastabilen austenitischen CrNi-Stählen. Habilitation: TU Bergakademie Freiberg; 2011

[18] Nikulin I, Sawaguchi T, Tsuzaki K. Effect of alloying composition on low-cycle fatigue properties and microstructure of Fe–30Mn–(6−x)Si–xAl TRIP/TWIP alloys. Materials Science and Engineering A. 2013;**587**:192-200. DOI: 10.1016/j.msea.2013.08.061

[19] Chen AY, Ruan HH, Wang J, Chan HL, Wang Q, Li Q, et al. The influence of strain rate on the microstructure transition of 304 stainless steel. Acta Materialia. 2011;**59**:3697-3709. DOI: 10.1016/j.actamat.2011.03.005

[20] Shen YF, Li XX, Sun X, Wang YD, Zuo L. Twinning and martensite in a 304 austenitic stainless steel. Materials Science and Engineering A. 2012;**552**:514-522. DOI: 10.1016/j.msea.2012.05.080

[21] Weidner A, Martin S, Klemm V, Martin U, Biermann H. Stacking faults in high-alloyed metastable austenitic cast steel observed by electron channelling contrast imaging. Scripta Materialia. 2011;**64**(6):513. DOI: 10.1016/j.scriptamat.2010.11.028

[22] Biermann H, Solarek J, Weidner A. SEM investigation of high-alloyed austenitic stainless cast steels with varying austenite stability at room temperature and 100°C. Steel Research International. 2012;**83**:512-520. DOI: 10.1002/srin.201100293

[23] Wendler M, Hauser M, Fabrichnaya O, Krüger L, Weiß A, Mola J. Thermal and deformation-induced phase transformation behavior of Fe–15Cr–3Mn–3Ni–0.1N–(0.05-0.25) C austenitic and austenitic–martensitic cast stainless steels. Materials Science and Engineering A. 2015;**645**:28-39. DOI: 10.1016/j.msea.2015.07.084

[24] Peter Palzer. Bestimmung der Stickstofflöslichkeit im schmelzflüssigen und festen Zustand bei hochlegierten CrMnNi-Stahlgusslegierungen mit anschließender Charakterisierung der Mikrostruktur sowie der mechanischen Eigenschaften. Diploma Thesis. TU Bergakademie Freiberg, 2012.

[25] Wendler M, Reichel B, Krüger L, Weiß A, Mola J. Influence of Carbon on the Microstructure and Mechanical Properties of Cast Austenitic Fe-19Cr-4Ni-3Mn-0.15N Steels. Hamburg, Germany: HNS 2014. p. 2014

[26] Hauser M, Wendler M, Ghosh Chowdhury S, Weiß A, Mola J. Quantification of deformation induced α'-martensite in Fe–19Cr–3Mn–4Ni–0·15C–0·15N austenitic steel by in situ magnetic measurements. Materials Science and Technology. 2015;**31**:1473-1478. DOI: 10.1179/1743284714Y.0000000731

[27] Wendler M, Mola J, Reichel B, Krüger L, Weiß A. Temperature dependence of properties in an Fe-19Cr-4Ni-3Mn-0.15N-0.2C austenitic cast steel. Aachen: HMnS 2014; 2014. p. 407-410

[28] Mola J, Ullrich C, Kuang B, Rahimi R, Huang Q, Rafaja D, et al. Austenitic nickel- and manganese-free Fe-15Cr-1Mo-0.4N-0.3C steel: Tensile behavior and deformation-induced processes between 298 K and 503 K (25 °C and 230 °C). Metallurgical and Materials Transactions A: Physical Metallurgy and Materials Science. 2017;**48**:1033-1052. DOI: 10.1007/s11661-017-3960-x

[29] Olson GB, Cohen M. Kinetics of strain-induced martensitic nucleation. Metallurgical Transactions A. 1975;**6**:791-795. DOI: 10.1007/BF02672301

[30] Martin S, Wolf S, Martin U, Krüger L. Influence of temperature on phase transformation and deformation mechanisms of cast CrMnNi-TRIP/TWIP steel. Solid State Phenomena. 2011;**172-174**:172-177. DOI: 10.4028/www.scientific.net/SSP.172-174.172

[31] Talonen J, Hänninen H. Formation of shear bands and strain-induced martensite during plastic deformation of metastable austenitic stainless steels. Acta Materialia. 2007;**55**(18):6108. DOI: 10.1016/j.actamat.2007.07.015

[32] Bogers AJ, Burgers WG. Partial dislocations on the {110} planes in the B.C.C. lattice and the transition of the F.C.C. into the B.C.C. lattice. Acta Metallurgica. 1964;**12**:255-261. DOI: 10.1016/0001-6160(64)90194-4

[33] Olson GB, Cohen M. A general mechanism of martensitic nucleation: Part II. FCC → BCC and other martensitic transformations. Metallurgical Transactions A. 1976;**7**:1905-1914

[34] Lee T-H, Ha H-Y, Kang J-Y, Moon J, Lee C-H, Park S-J. An intersecting-shear model for strain-induced martensitic transformation. Acta Materialia. 2013;**61**:7399-7410. DOI: 10.1016/j.actamat.2013.08.046

[35] Yang X-S, Sun S, X-L W, Ma E, Zhang T-Y. Dissecting the mechanism of martensitic transformation via atomic-scale observations. Scientific Reports. 2014;**4**:1-7. DOI: 10.1038/srep06141

[36] Kutschke T. Untersuchung zur deformationsinduzierten Martensit- und Zwillingsbildung im Gusszustand des Cr-Mn-Ni Stahls 16-7-6. Diploma Thesis. TU Bergakademie Freiberg; 2011

[37] Weiß A, Gutte H, Mola J. Contributions of ε and α' TRIP effects to the strength and ductility of AISI 304 (X5CrNi18-10) austenitic stainless steel. Metallurgical and Materials Transactions A: Physical Metallurgy and Materials Science. 2016;**47**:112-122. DOI: 10.1007/s11661-014-2726-y

[38] Hamada AS, Karjalainen LP, Misra RDK, Talonen J. Contribution of deformation mechanisms to strength and ductility in two Cr–Mn grade austenitic stainless steels. Materials Science and Engineering A. 2013;**559**:336-344. DOI: 10.1016/j.msea.2012.08.108

[39] Jahn A, Kovalev A, Weiß A, Wolf S, Krüger L, Scheller PR. Temperature depending influence of the martensite formation on the mechanical properties of high-alloyed Cr-Mn-Ni As-cast steels. Steel Research International. 2011;**82**:39-44. DOI: 10.1002/srin.201000228

[40] Kovalev A, Jahn A, Weiß A, Scheller PR. Characterization of the TRIP/TWIP effect in austenitic stainless steels using stress-temperature-transformation (STT) and deformation-temperature-transformation (DTT) diagrams. Steel Research International. 2011;**82**:45-50. DOI: 10.1002/srin.201000245

[41] Kovalev A, Jahn A, Weiß A, Wolf S, Scheller PR. Stress-temperature-transformation and deformation-temperature-transformation diagrams for an austenitic CrMnNi as-cast steel. Steel Research International. 2011;**82**:1101-1107. DOI: 10.1002/srin.201100065

[42] Kovalev A, Jahn A, Weiß A, Wolf S, Scheller PR. STT and DTT diagrams of austenitic Cr–Mn–Ni as-cast steels and crucial thermodynamic aspects of $\gamma \rightarrow \alpha'$ transformation. Steel Research International. 2012;**83**:576-583. DOI: 10.1002/srin.201100267

[43] Wendler M, Weiß A, Krüger L, Mola J, Franke A, Kovalev A, et al. Effect of manganese on microstructure and mechanical properties of cast high alloyed CrMnNi-N steels. Advanced Engineering Materials. 2013;**15**:558-565. DOI: 10.1002/adem.201200318

[44] Zhang W, Hu J. Effect of annealing temperature on transformation induced plasticity effect of a lean duplex stainless steel. Materials Characterization. 2013;**79**:37-42. DOI: 10.1016/j.matchar.2013.02.003

[45] Lecroisey F, Pineau A. Martensitic transformations induced by plastic deformation in the Fe-Ni-Cr-C system. Metallurgical and Materials Transactions B. 1972;**3**:391-400. DOI: 10.1007/BF02642042

[46] Olson GB, Cohen M. A general mechanism of martensitic nucleation: Part I. General concepts and the FCC → HCP transformation. Metallurgical Transactions A. 1976;**7**:1897-1904. DOI: 10.1007/BF02659822

[47] Cohen JB, Weertman J. A dislocation model for twinning in f.c.c. metals. Acta Metallurgica. 1963;**11**(8):996. DOI: 10.1016/0001-6160(63)90074-9

[48] Lewis N, Cieslak MJ, Savage WF. Microsegregation and eutectic ferrite-to-austenite transformation in primary austenite solidified CF-8M weld metals. Journal of Materials Science. 1987;**22**:2799-2810. DOI: 10.1007/BF01086474

[49] Martin S, Fabrichnaya O, Rafaja D. Prediction of the local deformation mechanisms in metastable austenitic steels from the local concentration of the main alloying elements. Materials Letters. 2015;**159**:484-488. DOI: 10.1016/j.matlet.2015.06.087

[50] Sakuma Y, Matlock DK, Krauss G. Intercritically annealed and isothermally transformed 0.15 Pct C steels containing 1.2 Pct Si-1.5 Pct Mn and 4 Pct Ni: Part II. effect of testing

temperature on stress-strain behavior and deformation-induced austenite transformation. Metallurgical Transactions A. 1992;**23**:1233-1241. DOI: 10.1007/BF02665054

[51] Feng W, Wu Z, Wang L, Speer JG. Effect of testing temperature on retained austenite stability of cold rolled CMnSi steels treated by quenching and partitioning process. Steel Research International. 2013;**84**:246-252. DOI: 10.1002/srin.201200129

[52] Byun TS, Hashimoto N, Farrell K. Temperature dependence of strain hardening and plastic instability behaviors in austenitic stainless steels. Acta Materialia. 2004;**52**:3889-3899. DOI: 10.1016/j.actamat.2004.05.003

[53] Shin HC, Ha TK, Chang YW. Kinetics of deformation induced martensitic transformation in a 304 stainless steel. Scripta Materialia. 2001;**45**:823-829. DOI: 10.1016/S1359-6462(01)01101-0

[54] Padilha AF, Rios PR. Decomposition of austenite in austenitic stainless steels. ISIJ International. 2002;**42**:325-337

[55] Bhadeshia HKDH, Honeycombe SR. 1 - Iron and its Interstitial Solid Solutions. In: Bhadeshia HKDH, Honeycombe SR, editors. Steels. 3rd ed. Oxford: Butterworth-Heinemann; 2006. p. 1-16

[56] Kiani-Rashid AR, Edmonds DV. Phase transformation study of aluminium-containing ductile cast irons by dilatometry. Materials Science and Engineering A. 2008;**481-482**:752-756. DOI: 10.1016/j.msea.2007.02.167

[57] Simmons JW. Overview: High-nitrogen alloying of stainless steels. Materials Science and Engineering A. 1996;**207**:159-169. DOI: 10.1016/0921-5093(95)09991-3

[58] Kim S-J, Lee T-H, Oh C-S. Effect of nitrogen on the deformation behaviour of high-nitrogen austenitic stainless steels. Steel Research International. 2009;**80**:467-472. DOI: 10.2374/SRI09SP033

[59] Lee T-H, Ha H-Y, Hwang B, Kim S-J, Shin E, Lee JW. Scale-bridging analysis on deformation behavior of high-nitrogen austenitic steels. Microscopy and Microanalysis. 2013;**19**:77-82. DOI: 10.1017/S1431927613012385

[60] Berns H, Gavriljuk V, Riedner S. High Interstitial Stainless Austenitic Steels. Berlin, Heidelberg: Springer Berlin Heidelberg; 2013

[61] Huang Q, Volkova O, Biermann H, Mola J. Tensile elongation of lean-alloy austenitic stainless steels: Transformation-induced plasticity versus planar glide. Materials Science and Technology. 2017;**0**:1-7. DOI: 10.1080/02670836.2016.1277091

Chapter 3

Investigation of Martensitic Transformation Induced by Cyclic Plastic Deformation in Austenitic Steels

Jerzy Kaleta, Przemysław Wiewiórski and
Wojciech Wiśniewski

Additional information is available at the end of the chapter

http://dx.doi.org/10.5772/intechopen.71408

Abstract

The main goal was to demonstrate the possibility of investigating martensitic transformation induced by plastic strain, especially including the kinetics of this transformation, using selected cross effects. It is commonly known that this type of transformation is a basic "mechanism" occurring in shape memory materials and metastable austenitic steels strengthened with martensite separations. The motivation behind the research was also to follow and visualise the transformation on line, during cyclic loading (fatigue process), without the necessity to use, for example, roentgenographic (destructive) or microscopic methods. The application of the magneto-mechanical effect (the Villari effect) and the thermomechanical effect (the Kelvin/Thomson effect) turned out to be particularly useful because they significantly change with martensite initiation and then accumulate in austenite. Therefore, the goal was to develop the non-destructive methods of investigating martensite transformation, which could then be used on real constructions made of metastable austenite steel. In the case of the magneto-mechanical method, the goal was to additionally visualise the magnetic field transformations along a sample in the function of a loading cycle and the index of this period. To achieve this, high-resolution phase maps were used, which also allowed image processing methods known from machinery visioning (MV) or digital image correlation (DIC) techniques to be used.

Keywords: martensite transformation induced by plastic strain, transformation kinetics, martensite initiation and accumulation, research methods, cross effects, Villari effect, Kelvin effect

1. Introduction and purpose of research

1.1. Significance of the research problem

Austenitic steels, due to their special mechanical, magnetic and technological properties, as well as corrosion resistance, are a material commonly used in the medical, automotive, aircraft, food, chemical, petrochemical and mining industries. The past 100 years of austenite steels have seen many examples of spectacular applications that justify today's position of this class of materials in economy and science. The issue has been widely discussed in numerous publications.

Martensite transformation that takes place in austenite steels significantly changes the properties of the input material. The appearance of martensite separations in the austenite matrix results in both positive consequences (e.g. strengthening of material) and negative ones. In other cases, such a change of structure is undesirable (e.g. a decrease in corrosion resistance, lower fatigue and fracture strength, the appearance of an undesirable magnetic phase).

There are two main types of martensite transformation: thermal transformation resulting from austenite supercooling at high speed and also strain-induced transformation resulting from the occurrence of critical plastic deformation. Martensite can also be obtained by the application of a strong magnetic field.

The martensite transformation induced by plastic deformation, including cyclical deformation, is a fundamental "mechanism" occurring in shape memory materials and metastable austenite steels strengthened with martensite separations.

The paper only focuses on the **martensite transformation induced by plastic strain** in metastable austenite steels, induced by cyclic plastic strain, and especially the methodology of investigating the phase transformation. It was considered especially important to develop efficient research methodology on the **kinetics of martensite transformation induced by cyclical plastic strain** using non-destructive methods. This is essential because, for example, the creation of constitutive and fatigue models for this class of materials [1–3] requires the estimation of the quantitative share of martensite as a function of the number of loading cycles.

On the other hand, investigating the kinetics is not easy because in each mechanical loading cycle there is a martensite growth, which means that the material which is initially a single-phase one (austenite) becomes a two-phase material (austenite + martensite). Additionally, the austenite-martensite quantitative ration is systematically changed. It should also be emphasised that the research results encompassing mechanical values (stress, strain, the area of the hysteresis loop, etc.) only allow the behaviour of a two-phase material to be cyclically analysed, without a possibility to obtain more precise information about martensite.

There are a few methods of investigating martensite transformation, amongst others, roentgen diffraction [4–6], neutron diffraction [7], microscopic measurements [6, 8], density [4], hardness measurements [4] as well as magnetic measurements [4–6, 8–12]. There is still a need, however, for non-destructive methods, which would be useful in both laboratory and industrial conditions. This is why, as is stated below, new methods using selected physical cross effects can turn out to be especially effective.

http://dx.doi.org/10.5772/intechopen.71408

Transformation in two- and three-dimensional objects. It is assumed that the martensite transformation induced by plastic strain (which is also the consequence of supercooling) occurs differently in massive objects in which all three dimensions are of the same order (the so-called 3D objects) and also in two-phase objects such as foils (2D objects). In 3D objects, the transformation process is initiated and developed in a different way in the external layer, known as the skin, compared to deeper layers belonging to the so-called core. The other material, that is, 2D, can be treated as the skin, which is a more uniform object.

1.2. Research objective

The main goal of the research was to indicate the possibility of investigating the martensite transformation induced by plastic strain using selected cross effects, including mainly the kinetics of this transformation. The application of the magneto-mechanical and thermomechanical effects was considered particularly useful, because, as it was shown, they change significantly with the initiation and later accumulation of martensite in austenite. Hence, the goal was the development of non-destructive research methods of martensite transformation, which could be used both in laboratory conditions and on real constructions made of metastable austenite steel.

Key issues:

- Selection of cross effects and proving their usefulness in the online research on martensite transformation
- Construction of dedicated measurement apparatus for laboratory research which could also be used in industrial conditions
- Verification of the method in laboratory conditions

2. Cross effects used to investigate the phase transformation

In the research on the kinetics of martensite transformation induced by plastic strain presented below, the use of two cross effects, namely, the Villari effect and the Kelvin (Thomson) effect, was proposed.

Villari effect application. The phenomenon of inverse magnetostriction, also called the Villari effect, is particularly useful in the research on transformation. The essence of the effect was described in numerous earlier publications, also written by the authors of this article [13, 14]. It utilises the susceptibility of the effect of the occurrence of the ferromagnetic phase, that is, martensite α'. Thus, it is possible to easily detect the moment of transformation initiation after exceeding the cyclical border of plasticity in metastable austenite. It is also possible to assess the martensite phase growth, depending on the level of loading and the number of cycles. The local character of the measurement (due to the small size of a sensor, e.g. a magnetoresistor) allows a particular way of scanning the investigated surface for the purpose of finding a ferromagnetic phase in the "background" of the dominant austenite phase to be conducted.

It was assumed that it would be possible to indirectly investigate the growth of martensite in austenite by analysing the waveform of a selected magnetic component and by "calibrating" it in a certain way. The magnetic signal, which is a representative for martensite separation and accumulation, will be indicated at a later stage.

Kelvin effect application. An example of another effective research method is the application of the thermo-elastic phenomenon, also called the Thomson effect or the Kelvin effect. The effect used the fact that in adiabatic conditions—in terms of elastic strains—metals behave like gas. This means that during tension they are cooled down and during compression they are heated up. In a triaxial stress state, the change of temperature ΔT is also directly proportional to the sum of changes in the normal stresses ($\Delta\sigma_{xx}$, $\Delta\sigma_{yy}$, $\Delta\sigma_{zz}$) of the linear coefficient of thermal expansion α and absolute body temperature T_0 and indirectly proportional to density ϱ and the specific heat c of a body; hence,

$$\Delta T = -\frac{T_0 \cdot \alpha}{\rho \cdot c}\left(\Delta\sigma_{xx} + \Delta\sigma_{yy} + \Delta\sigma_{zz}\right) = -k\sum_i \Delta\sigma_i \qquad (1)$$

For over 150 years (from 1853) [15], it was thought that this law was well documented experimentally. However, not so long ago, it was found that there was an interesting deviation from the Kelvin model in the case of ferromagnetic materials [16]. **Figure 1**, showing waveforms $\sigma(\xi)$ and $\Delta T(\xi)$ [17], presents clear evidence to this.

Sinusoidal stress waveforms—according to Kelvin's law—should correspond within counterphase sinusoidal signals of temperature growth ΔT. This fact actually occurs until a certain value of loading is reached, before which there is only a paramagnet, that is, austenite. The appearance of even a small amount of martensite α' (ferromagnetic phase) results in the occurrence of a certain kind of saddle in waveform $\Delta T(\xi)$ or, in other words, higher harmonic components.

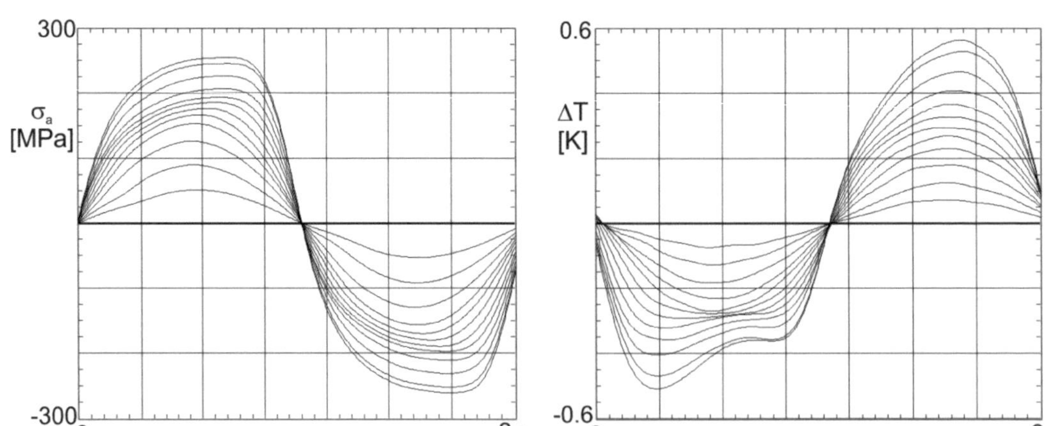

Figure 1. Stress waveforms and the temperature growth ΔT waveforms corresponding with them, which offer evidence for the occurrence of the martensite transformation induced by plastic strain [17].

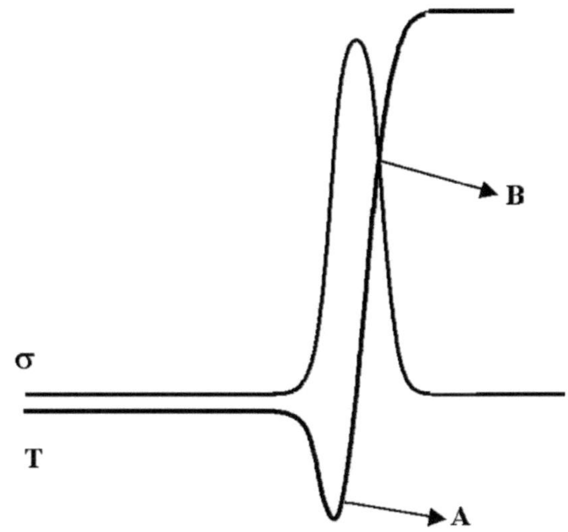

Figure 2. Thomson's law: range of action.

This phenomenon was discovered [16] when temperature was measured with very thin thermocouples (OMEGA type J 0.001″ ~ 25 μm), which have low thermal inertia and allowed higher harmonic components to be registered.

It should be emphasised that the similarity in the behaviour of metals and gases ceases to exist when plastic strain occurs in metal, which is presented in **Figure 2**. It can be observed (**Figure 2**) that temperature decreases with an increase in stress. The appearance of plastic strains in a material is followed by a considerable increase in temperature (point A). At point B, the material returns to the range of plastic strains, and a decrease in load is accompanied by a further growth in temperature because of the occurrence of compression. However, it follows Thomson's law. The temperature change between points A and B confirms the appearance of plastic strains in the material, and the rapid nature of this change is significant enough to be easily observed.

3. Evaluation of martensitic transformation induced by plastic deformation in massive specimens

3.1. Research object and measurement setup

The object of the research was to investigate cylindrical samples made of AISI 304 austenite steel. A sample is presented in **Figure 3**, and the chemical composition and strength properties are given in **Table 1**. Before the tests the material was heat treated, and the samples were held at a temperature of 1050°C for 35 minutes and were then quickly cooled in water. The goal of the thermal treatment was to obtain a single-phase material with austenite structure.

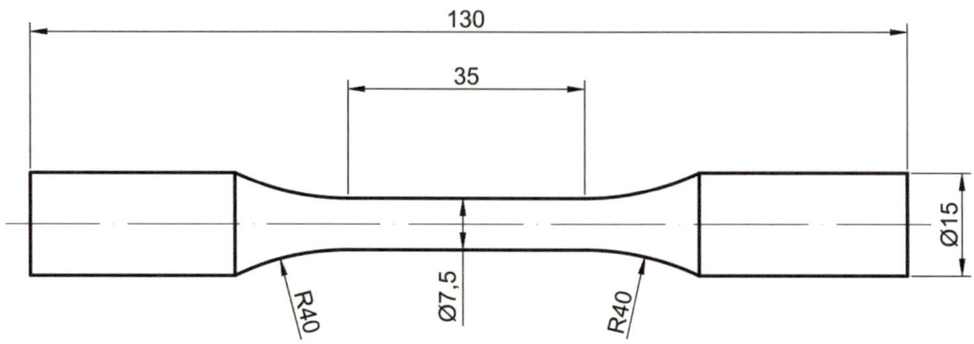

Figure 3. Drawing of the sample [18].

Material	Chemical composition								Strength properties		
	C	Si	Mn	P	S	Cr	Ni	Ti	R_e	R_m	A_5
	%	%	%	%	%	%	%	%	MPa	MPa	%
AISI 304	0.018	0.297	1.402	0.032	0.026	18.347	8.757	0.006	593	752	34

Table 1. Chemical composition and strength properties of AISI 304 steel [18].

The measurement system was composed of a testing machine MTS 810, computer, extensometer and a magnetic field measuring device (**Figure 4**). The testing machine was controlled by a FlexTest GT controller and MultiPurpose TestWare Software, which allowed mechanical signals (stress σ(t) and strain ε(t)) and magnetic signals (magnetic field strength H(t)) to be registered. Magnetic field strength was measured using an original system constructed by the authors' team. In the system, four magnetoresistors KMZ 10B made by Philips (NXT) were used as sensors, and they were arranged in a line according to the sample measurement

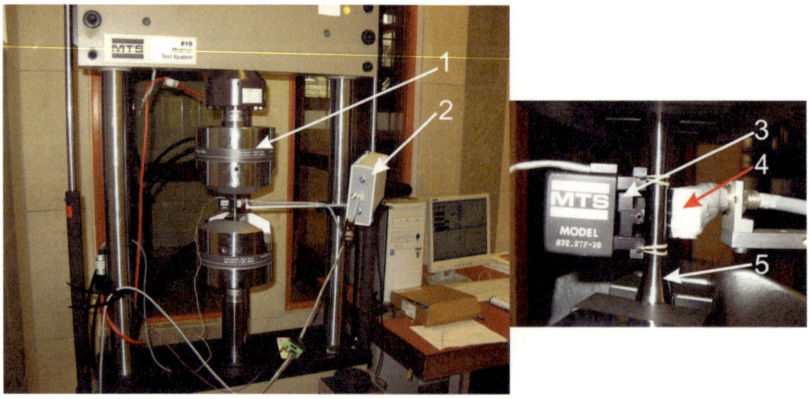

Figure 4. Measurement setup: (1) hydraulic pulsator MTS 810, (2) magnetic field measurement system, (3) extensometer, (4) magnetoresistor and (5) sample [18].

http://dx.doi.org/10.5772/intechopen.71408

length. The used sensors are characterised by good measurement properties and a possibility to reduce the background field (the Earth's magnetic field and the field generated by steel objects surrounding the sensor). The accurate positioning of sensors and the method of conducting measurements are presented later in the paper.

The prepared measurement set was used to conduct low-cycle fatigue tests. The fatigue process was controlled by the total strain amplitude. The research was conducted for four thresholds, ε_a = 0.003, 0.0035, 0.004 and 0.005 mm/mm, respectively. Five samples were tested at each of the thresholds. The process was continued until a sample was broken. The load spectrum had a sinusoidal character (R = −1), and the frequency was 0.2 Hz. The selected value of frequency counteracted the temperature growth resulting from material deformation, which prevented inverse transformation, That is, martensite $\alpha' \rightarrow$ austenite γ.

Figure 5 presents a low-cycle fragment of the Wöhler diagram. **Figure 6** presents results for a selected sample for input function ε_a = 0.0035 mm/mm. **Figure 6** presents the change of the area of the mechanical hysteresis loop and selected loops from the whole lifecycle of the sample. Moreover, some symbolic points corresponding with selected loops are marked in the figure. It is visible that the mechanical loop area decreases with the number of cycles, which confirms the strengthening of material when experimental control with total strain ε_a is used. Before the tested sample broke, it had been loaded with N = 24,723 cycles.

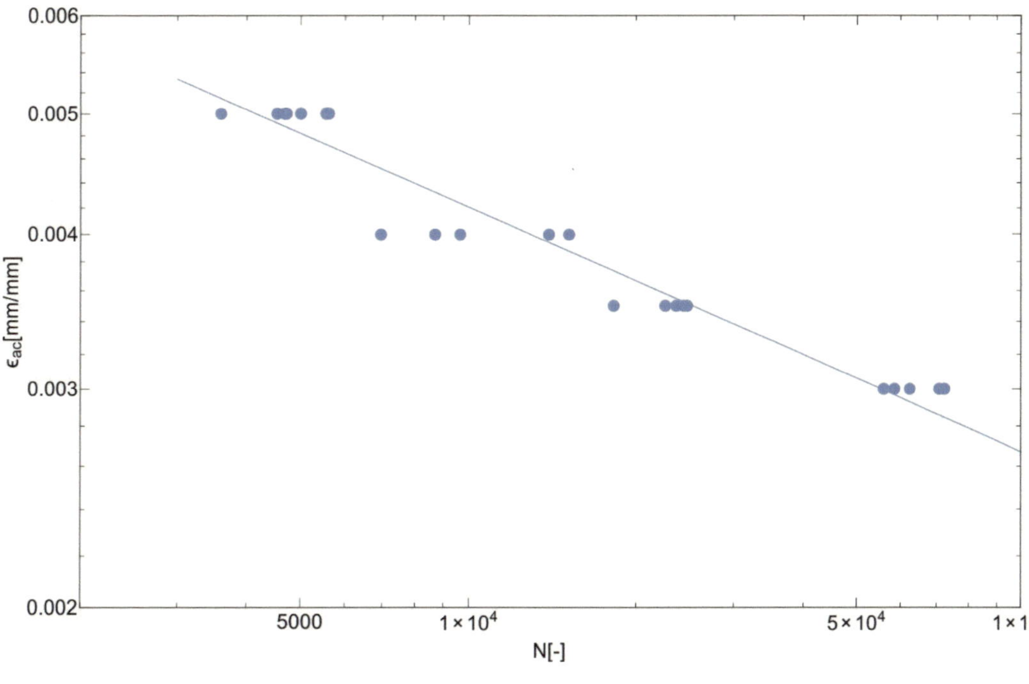

Figure 5. Low-cycle fragment of the Wöhler diagram [19].

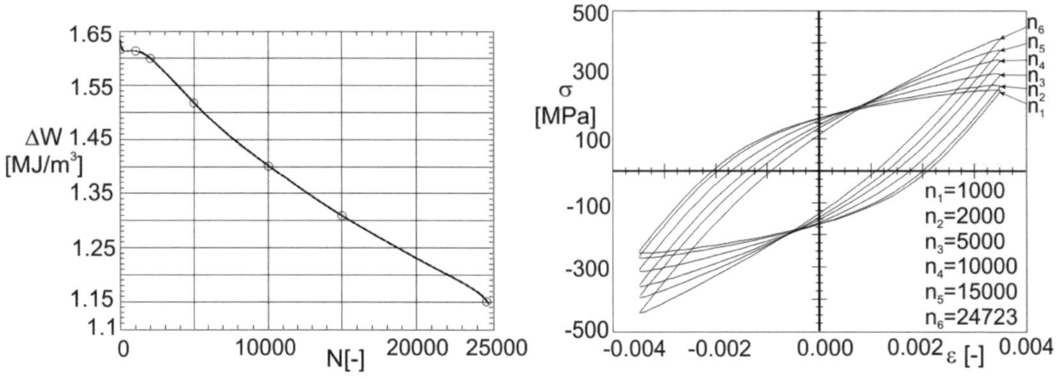

Figure 6. Area of the hysteresis loop with selected loops [19].

3.2. Methodology of magnetic measurements

3.2.1. Preconditioning of magnetic measurements

Contactless measurement of magnetic field strength in the conditions existing during loading samples in a testing machine is a significant technical challenge [18, 20]. The measurements were conducted with a passive contactless method in which magnetoresistors were used as quasi-pointwise magnetic field sensors [13]. Before the experiment started, there were a number of conditions which had to be taken into account. The most important of them are briefly discussed below:

a. Ensuring of non-magnetic surroundings around the tested sample is an ideal solution for passive methods. However, it is practically impossible due to the presence of a magnetic field. On the other hand, in the case of testing the separation of a martensite phase in paramagnetic austenite steels, the measurement signal is one order of magnitude smaller than the Earth's magnetic field [21].

b. It is hard to eliminate the so-called 50 Hz magnetic field of electrical appliances.

c. A hydraulic pulsator is made of numerous ferromagnetic parts that move during its operation. This means that the pulsator itself is a magneto-mechanical system. It was assumed that the introduction of interference in the machine system is pointless due to its ability to influence the correctness of performed fatigue loads. Moreover, irrespective of this, ensuring a zero magnetic field was unrealistic.

d. The strongest magnetic field is related to the presence of strongly ferromagnetic wedges made of forged tool steel. Such a magnetic circuit is characterised by the opposite polarisation of the system of wedges along a sample, which is essential in the design of a measurement setup. The wedges have a knurled surface, which causes the permissible/acceptable collapse of the surface of a sample. An example of the influence of the knurled surface of the wedges is presented in a magnetic "picture" of an austenite steel sample that was taken with the author's original magnetovision camera (**Figure 7**) [21, 22]. There

http://dx.doi.org/10.5772/intechopen.71408

are visible traces of the collapse caused by the knurled surface and the location of technological cutting of the sample. Hence, it can be observed that even only compressing the sample in the wedges leads to the local initiation of a magnetically strong martensite phase.

e. The location of quasi-pointwise magnetic sensors with reference to the sample is important in the analysis of the magnetic circuit of the sample and the testing machine. The application of magnetoresistors is optimal due to the linear character of the response in terms of measurements and the very high sensitivity with relatively low noise. In the system of a typical Wheatstone bridge, these sensors are very sensitive measurement tools. For instance, when the signal is magnified about 1000 times, it is possible to record the process, through a wall from a distance of 10 m, the process of parking a car. In addition to this, these are wideband devices that can operate between a DC and AC signals at the frequency of a few MHz [23].

3.2.2. Position of sensors and the magnetic circuit of the sample and testing machine

The magnetic circuit of the sample and the testing machine, taking into consideration the above conditions, is presented in **Figure 8A**. Austenite sample **1** is fixed symmetrically in the strongly ferromagnetic wedges of the testing machine (top **2A** and bottom **2B)**.

Two measurement devices are essential in mechanical measurements: extensometer **3** and force transducer **4**. The extensometer **3** with a specific measurement base (25 mm) is fixed symmetrically on the sample using special non-magnetic grips. The extensometer contains ferromagnetic parts in its construction (the so-called knives), and a small current flows through the device. However, it was assumed that placing the extensometer on the other side of the sample would not cause it to be a source of a significant interference field.

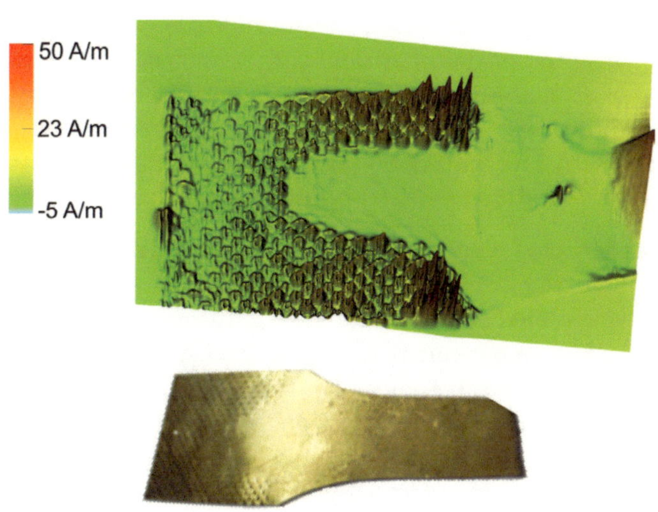

Figure 7. Visualisation of the magnetic field of the knurled sample AISI 304 compressed in the testing machine [21].

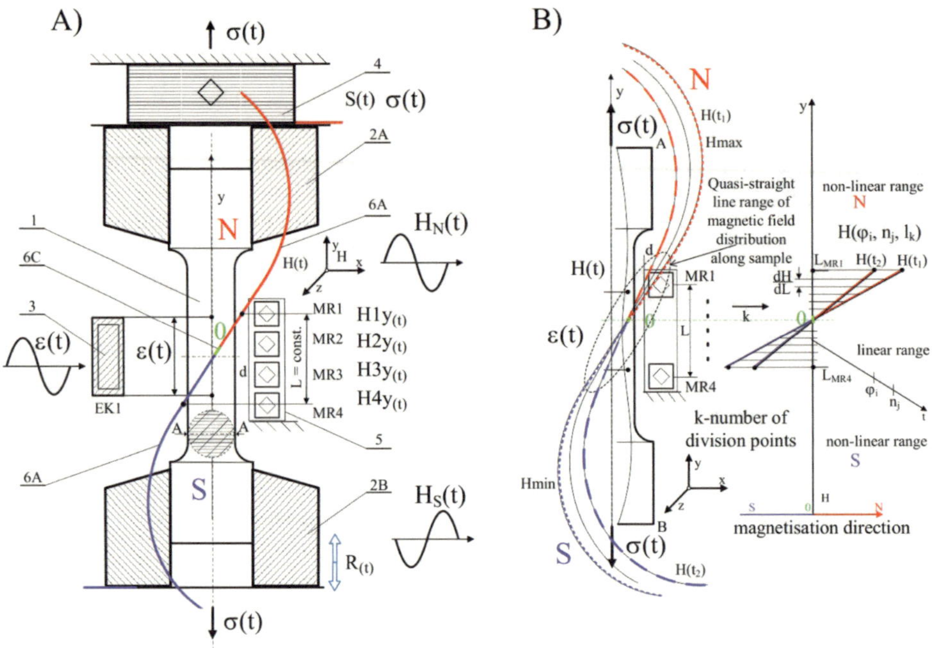

Figure 8. Scheme of magnetic field propagation in the magnetic circuit related to the testing machine: (A) measurement structure with the lines of the predicted propagation of magnetic field along the sample length, (B) scheme of the change of magnetic field propagation under the influence of a mechanical load. Subassemblies presented in the figure: (1) massive cylindrical sample; (2) testing machine wedges: 2A and 2B (top and bottom); (3) extensometer –(strain measurement value ε(t) and piston position change R(t)); (4) force measurement transducer (stress measurement value σ(t), force S(t)); (5) magnetic field measurement head (magnetic field strength measurement value H(t) at particular points with regard to zero magnetic field); (6) magnetic field distribution line: 6A and 6B (magnetic poles, respectively, N and S) and 6C (zero magnetic field location, the so-called magnetic zero line). Note: The sinusoidal character of signal H(t) is a measured idealisation, which facilitates the explanation of assumptions; in reality this signal is complex and contains higher component harmonics.

The force transducer **4** is made of ferromagnetic materials; however, it is located at a significant distance from the sample on one of its sides with reference to the centre. Mechanical load was the application of a sinusoidal strain spectrum with a fixed value until sample destruction. The force signal **S**(t) was recorded along with strain changes **ε**(t) and recalculated to stress **σ**(t). The control and data acquisition system of the hydraulic pulsator also ensured the measurement of piston shift **R**(t).

After annealing with accepted zero (residual) content of martensite, sample **1** is fixed in the strongly ferromagnetic wedges of the testing machine, top **2A** and bottom **2B**. Two measuring devices are essential in mechanical measurements: extensometer **3** and force transducer **4**. Extensometer **3**, with a known measurement base, is fixed symmetrically on the sample using special non-magnetic grips. The extensometer has some parts made of ferromagnetic materials in its construction (e.g. the so-called knife edges), and a small current flows through the device. However, it was assumed that placing the extensometer on the other side of the sample would not cause it to be a source of a significant interference in magnetic background field. The key issue was preventing the possibility of the extensometer knife edges from sliding during the many hours of loading attempts. The force transducer **4** is made of ferromagnetic materials; however, it is located at a larger distance from the sample on one of its sides with reference to the centre.

Taking into account the magnetic circuit of the testing machine and the sample, as well as the sensitive axis of magnetoresistors (along the sample, i.e. the y axis) distributed opposite the extensometer, the distribution of the magnetic field takes the N-0-S form, where "0" is the zero value of the field between the opposite poles N-S. N-0-S divides the sample into two influence zones of magnetic poles in which the zero line does not necessarily have to overlap with the sample axis of the symmetry axis. The measurement of the magnetic field is conducted in a contactless way at a small distance d (usually 1 mm) from the created surface of the cylindrical sample. The aspired goal is to locate the magnetic zero between sensors MR1 and MR4 and also between the extensometer knife edges. Due to the fact that the austenite steel sample is initially paramagnetic, the position of the zero line will be the only information that can be obtained from magnetic sensor indications. Only the separation of martensite will cause changes in signal value growth. Thus, the task to be completed by magnetic sensors was the online monitoring of the changes of the zero line position in particular loading cycles.

Due to the fact that the austenite steel sample is initially paramagnetic, the position of the zero line will be the only information that can be obtained from magnetic sensor indications. Only the separation of martensite will cause changes in signal value growth. On the other hand, fixing only one sensor on any part of the cylindrical sample will mean that measurements are taken at only one of the poles or will be possibly made in the not-very-advantageous area influenced by the magnetic zero line. The location of the sensor near the magnetic zero line may, in turn, cause its low indications.

Thus, it is more advantageous to design the measurement head **5** in such a way that there are two sensors distant from each other—there should be a small, constant distance "L" between them along the y axis, but simultaneously they must be located on both sides of the magnetic zero line. However, it is better to use a larger number of magnetoresistors. The reason for this is that internal sensors can follow the changes of the position of the zero line in particular cycles. Another solution is mapping the created cylindrical surface using a magnetic scanner (magnetovisual) that is synchronous with regard to the mechanical measurement value.

Figure 8B symbolically shows the course of magnetic field distribution changes along the sample related to the location of magnetic poles. Under the influence of a mechanical load, the sample is deformed. As a result of the Villari effect, a small change appears in the sample in the magnetic field distribution and is accompanied by the change of the zero line position on the sample. Such a process was described as the disturbance of the "magnetostatic balance" (in an ideal case, when martensite separations were uniform in each part of the sample micro-volume, the location of the magnetic field zero line would not change). This means that a change in the "magnetostatic balance" results solely from the presence of the martensite phase located randomly on the sample. The sensors conduct a certain kind of homogenisation by averaging their measurement surface. In consequence, the disturbance of the "magnetostatic balance" in the presence of a directed magnetic field can be treated as the measure of the quantitative content of martensite in austenite. In the next step, the key stage in the procedure of determining the amount of martensite in online monitoring of its content is obtaining the maps of zero line position changes.

The key assumption adopted below is that the linear range field changes along the sample near the zero line. Under the influence of a mechanical load, only the inclination of straight lines will change (apart from the earlier discussed position of the magnetic zero line) (**Figure 8B**).

The defined characteristic of the flux distribution allowed high-geometric-resolution maps showing the fluctuation of the magnetic field along the cylindrical sample to be prepared. These maps depend on the number of the load cycle, the index of this period (i.e. the angle in the period in the range of 0–360°) and the value of strain.

3.2.3. *Methodology of magnetic measurements and their interpretation*

The main task performed by the acquisition system of measurement signals, using MTS FlexTest GT, was recording the following variables in time:

1. Strain $\varepsilon(t)$
2. Force S(t), used to determine stress $\sigma(t)$
3. Position of bottom grip piston R(t)
4. Magnetic field strength H(t) in the form of four waveforms, $H_1(t)$, $H_2(t)$, $H_3(t)$ and $H_4(t)$, respectively, obtained from a four-element head with sensors marked as MR1, MR2, MR3 and MR4

Due to the fact that magnetic measurements constitute a significant, technical challenge, the conducted research methodology described below presents the essential characteristics of the measurement head. Magnetic field acquisition is performed for sensors located on two opposite magnetic poles in the sample axis. The distance between the extreme sensors is small, typically 20 mm, which makes up 1/2 of the sample measurement part. The distance between the extremes of the magnetic field located in the poles, between the wedges, is at least five times bigger than the distance between two extreme sensors. The propagation of the magnetic field along the Y-axis in the place that is close to the magnetic zero can be interpreted as the zone of magnetic field linear increments with regard to sample length. The values of extreme sensors are not identical due to the fact that the position of the zero line is not known.

The magnetic field measurement head consists of four KMZ10 type magnetoresistors. Additionally, it is possible to make a head dedicated to various ranges of the geomagnetic field: for very weak magnetic fields of up to 200 A/m (KMZ10A, KMZ51), for medium ones of 1000 A/m (KMZ10B) and for strong ones of 10,000 A/m (KMZ10C).

Sensors from the KMZ10 family are standard magnetoresistor devices offered by NXP (previously Philips). They are characterised by a relatively big measurement area, thanks to which the level of value averaging per unit of length is significant. This, however, does not adversely influence measurements and is actually beneficial because the sensors are not sensitive to local geometric anomalies, that is, roughness, scratches and local martensite separations. These signals are more representative for the whole sample. Moreover, the ensured magnetic polarisation counteracts local magnetic anomalies, and as a result, sensor indications should be treated as "global" with regard to the magnetic zero line.

The process of setting sample load cycles was also conducted by using real-time control of the position R(t) of the bottom grip piston (when the top grip is immobile) in which the indications of the extensometer follow the profile

$$\varepsilon(t) = \varepsilon_{max} \cdot \sin(\omega t), \tag{2}$$

at frequency f = 0.2 Hz.

The fatigue process was registered until the sample broke. Throughout the whole time of the measurements, the sample was not elongated, which is essential with respect to immobile magnetic sensors.

The sampling of all measurement channels was conducted simultaneously at sampling frequency:

$$F_P = 360 \cdot f = 72\ \text{Hz}, \tag{3}$$

The synchronising period of the strain signal ε(t) was divided into i = 360 equal time sections (period index i = 0,…,359) with consideration for the j'th loading cycle n. Therefore, strains ε(t) were obtained as the functions of two variables φ_i and n_j connected with the period and the load index according to the following rule:

$$\varepsilon\left(\varphi_i, n_j\right) = \varepsilon_{max} \cdot \sin\left(\frac{j \cdot 2\pi}{360} \cdot i\right), \tag{4}$$

The representation of a given value identical with ε, σ and H can be analysed in the linear time increment (the so-called RAW, unprocessed data—an oscilloscopic course with a long-time basis) or with regard to the index i = 0 of the forcing value, in this case ε. The model course of strain $\varepsilon(\varphi_i, n_j)$ contains up-to-date information about the period index φ_i (i = 0,…,360). The course in time A(t) contained n_j cycles, where "j" is the indicator of the load cycle period completed with breaking a sample. All measurement values (ε, σ, H_1–H_4) were synchronised with reference to index i = 0 of the strain signal ε. The method of analysing the real signal waveforms of the magnetic field strength H(t) as RAW is shown in **Figure 9** on the example of four waveforms from the magnetoresistors (MR1, MR2, MR3 and MR4, respectively) that were fixed along the sample at equal distances from one another and in accordance with the scheme presented in **Figure 8**.

It should be noted that the signals in the counterphase from the MR1 and MR4 sensors are strictly correlated with semi-cycles of compression (ε_S)—tension (ε_R). The analysis of magnetic field $H(n_j)_{(\varphi i)}$, the maximum values for MR1 and the minimum values for MR4 refer to the same tension of semi-cycle; hence, all compression values (ε_S) are in the waveforms in **Figure 9**.

The analysis of registered values, when only on the basis of their RAW-type time waveforms, is insufficient. It was necessary to develop methodology that would allow access to each registered value. The key aspect is the strain level $\varepsilon(n_j)\varphi_i$ for each synchronised measurement value in any load cycle until a sample is broken. This means that there is a possibility, due to the images being consistent with the linear increment of period index φ_i that is identified with the constant strain level ε, to build and further analyse the results using high-information-density graphic methods. Here, techniques known from the development of magnetovision,

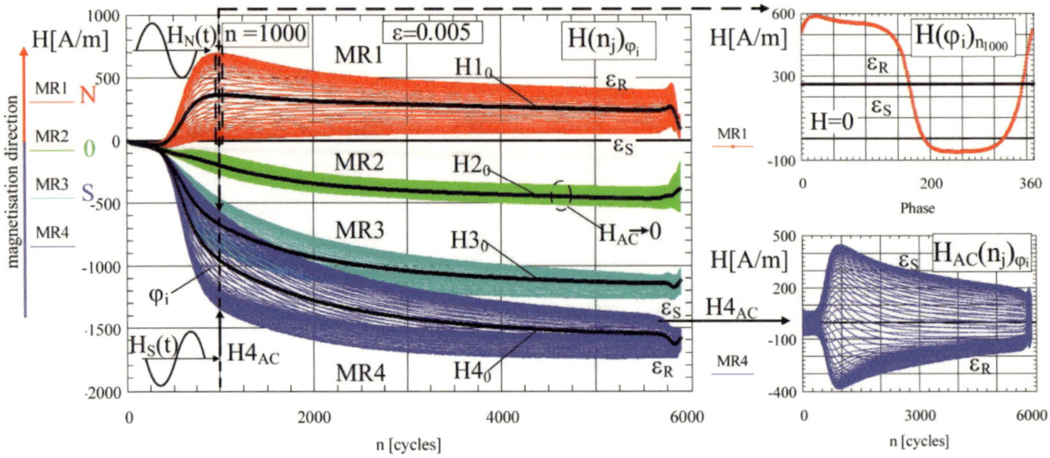

Figure 9. Example of the analysis of waveforms from the four magnetoresistors MR1, MR2, MR3 and MR4 located in accordance with **Figure 8**.

developed for many years by the authors of [21, 22], turn out to be especially useful. Recording values which are time synchronised with one another allows the drawing of maps and tomograms based on period index φ_i or load cycle n_j, defined as phase maps (2D) and tomograms (3D). It resembles the decomposition of an analogue TV signal transmitted as a time waveform to an image form in which a frame is a kind of two-dimensional map [24]. Additionally, the term phase means here that the maps are mainly used to determine phase shifts (phase-sensitive detector) with reference to the model that is the strain map $\varepsilon(\varphi_i, n_j)$. The maps can comprise of even several millions of points, which is an essential parameter in typical data analysis as it was shown charts [25]. The assumed map size usually has the dimensions of table 2D 360x j_{max}. In this way it is possible to conduct the imaging of various physical values, mapped with reference to the increment of a given parameter [26–28]. In the case of the martensite transformation, the main focus was on strain, stress and magnetic field maps.

Figure 10 presents the phase maps $H(\varphi_i, n_j)$ for the MR1 and MR4 sensors of a sample loaded with a profile with the maximum amplitude of total strain $\varepsilon_a = 0.005$. On the map level, lines mark the ranges of similar values of a magnetic field. The cross sections along indexes (i, j) allow waveforms as cycle periods, and also changes occurring under the influence of the fatigue process, to be obtained. **Figure 10** presents selected period waveforms for load cycle index j = 100 (1200th cycle of sample loading) with marked maximum and zero values (i = 80, i = 170, i = 260). The changes of a magnetic field under the influence of the fatigue process are presented for index i = 20.

The results in the form of maps are the basic tool in the cognition of the martensite transformation phenomenon and also an interface that allows image processing methods known from machinery visioning (MV) or DIC techniques [29] to be used.

The key assumption made by the authors is the linear range of field changes along the sample near the zero line. Under the influence of a mechanical load, only the inclination of straight lines along the sample, related to the propagation of the field, will change, as shown in **Figure 8B**.

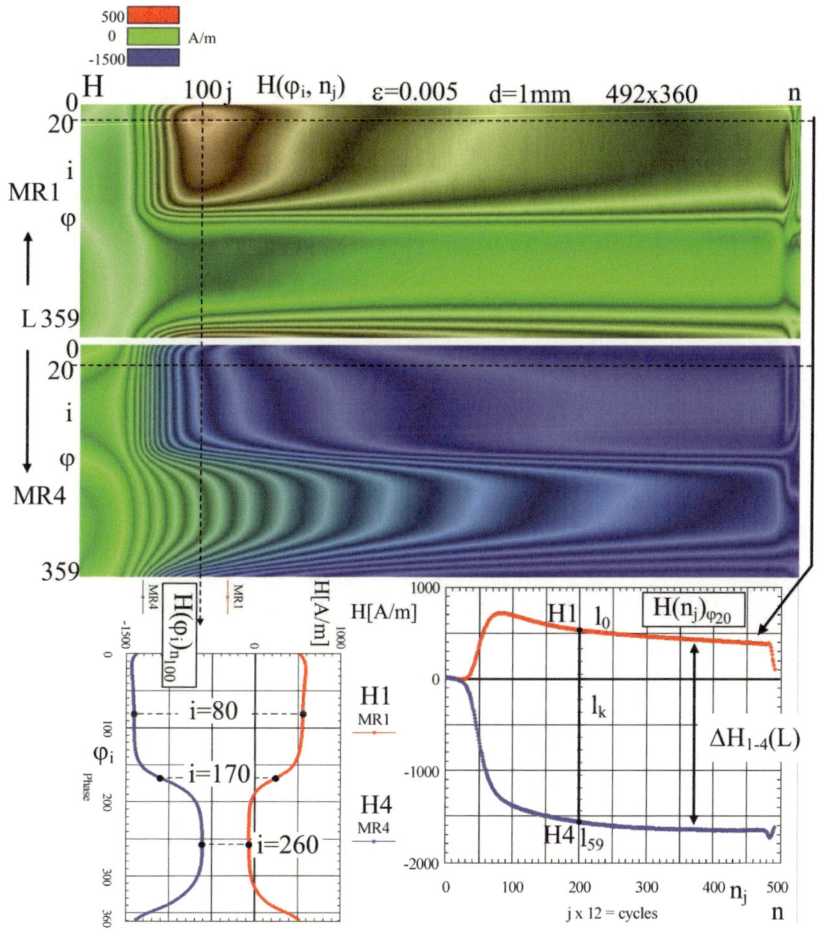

Figure 10. Phase maps for MR1 and MR4 sensors for the sample loaded with profile $\varepsilon_a = 0.005$.

Thanks to this assumption, it is possible to develop high-geometric-resolution maps that show the changes of a magnetic field along a cylindrical sample depending on a given load cycle and its period index (the angle in period in the range of 0–360°), related to a current strain. The H maps (φ_i, n_j, l_k) obtained for sample $\varepsilon_a = 0.005$ are presented in **Figure 11**. The level lines mark the ranges of similar values of a magnetic field for the same angle in period (phase index).

It was said earlier that the position of a sensor on the sample is essential because of the distance from the unknown position of the zero line. Any attempt to interpret a signal from a single sensor is influenced by its position with reference to the zero line of a magnetic field [30]. For the purpose of gaining independence of the zero line position, the operation of the determination of the increment of magnetic field strength depending on sample length H′ must be conducted:

$$H'(\varphi_i, n_j) = \frac{dH(\varphi_i, n_j, l_k)}{\mathrm{dl}}, \tag{5}$$

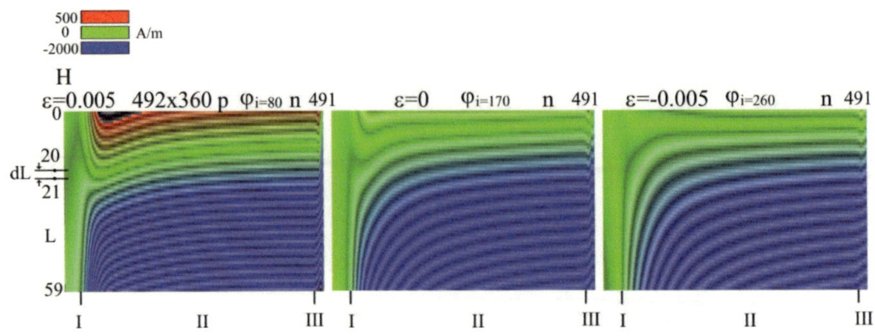

Figure 11. Magnetic field distribution maps used to determine zero line (H = 0) position changes under the influence of the number of cycles with regard to a set phase index.

The method allows for the consequences of some imperfections in the course of the fatigue process to be compensated, for example, minor sliding of extensometer knife edges or minimum sample elongation. **Figure 12** presents waveform H' with marked envelopes of maximum values, H'_{max} and H'_{min}, respectively. **Figure 13** presents a phase map with corresponding cross sections. Special attention should be drawn to the waveform of averaged value H'_0 (n_j), which is a characteristic value for the whole sample and depends only on the load cycle. Hence, all measurement results in the form of individual sensor waveforms, or maps can be reduced to the average value of magnetic field increment along path H'_0. Waveforms H'_0 (n) characterise the accumulation of martensite emitted in a sample during a fatigue process, with a set level of strain amplitude ε_a.

3.3. Application of the magnetic field increment method in martensite content assessment

Average magnetic field increment per unit length of the measurement part of sample H' was assumed to be the magnetic measure of the martensite phase content in a sample. The strain

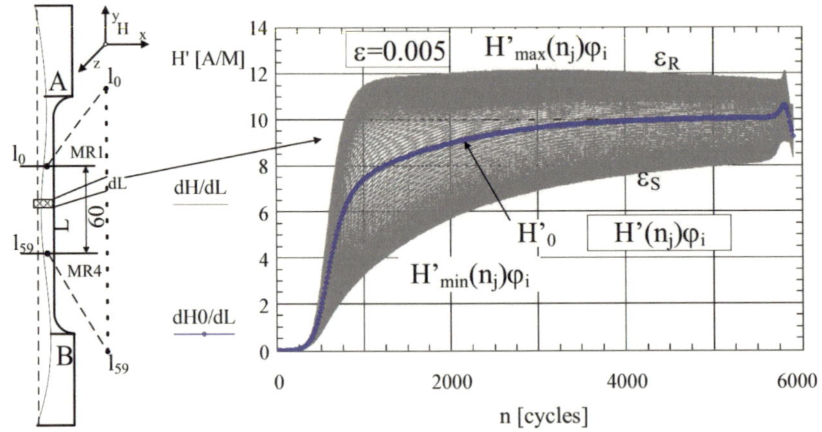

Figure 12. Waveforms of changes H'(n) depending on load cycle for set ε_a.

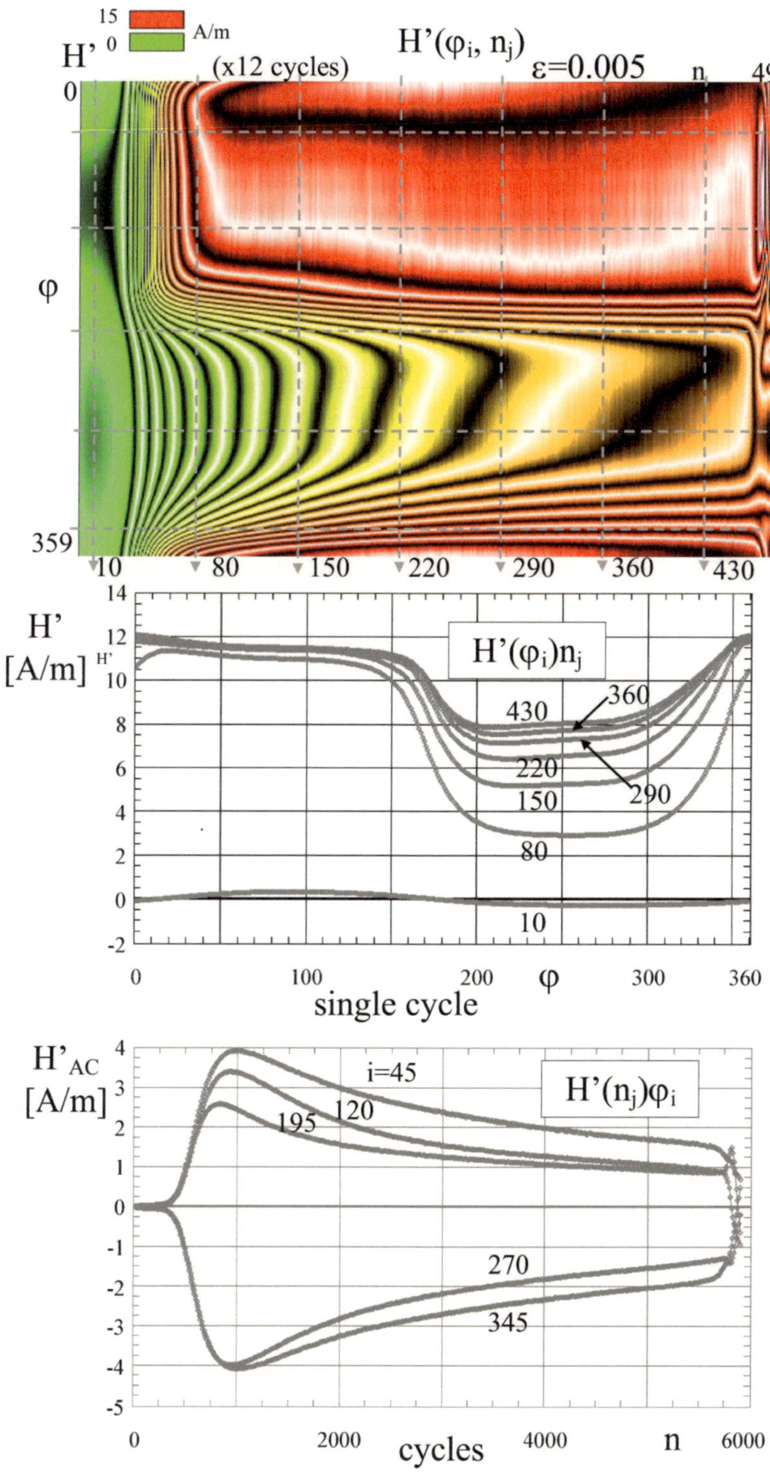

Figure 13. Map of changes in the increment of magnetic field H′ synchronised in period φ and load cycles for set ε_a.

level $\varepsilon_a = 0.005$ was the maximum level at which sample geometry did not change until the sample broke. The reason for this is that in earlier research it was shown that for $\varepsilon_a = 0.0055$, buckling took place very quickly. As a result, the value of $\varepsilon_a = 0.005$ was assumed to be the critical ε_k to which the results would refer, especially when the high repeatability of results for mechanical and magnetic values was detected.

The analysis of these waveforms allowed the so-called maximum amount of martensite V_H with regard to the highest strain values ($\varepsilon = 0.005$) to be determined:

$$V_H = \frac{H_0'(n_j)(\varepsilon_a)}{H_0'(n_j)(\varepsilon_k)} \cdot 100\%, \tag{6}$$

Figure 14 shows waveforms $H_0'(n_j)$, depending on the strain level, for three samples of various total strain levels with value V_H. The error in the results for the method is connected with the quotient of the critical total strain and the strain at which the sample buckles, in this case $\delta V_M = \frac{0.005}{0.0055}$, i.e. about 10%.

In **Figure 14**, value V_H is also marked and is related to the number of sample loading cycles before the sample breaks and the increment of magnetic field H_0', which is the measure of accumulated martensite content. The next step will be scaling values V_H in the same way as the accumulated martensite content in a fatigue process using optical microscopy or roentgenographic methods. The analysis of subsequent values $H_0'(n_j)$ with regard to the load cycle index allows the determination of the martensite increment speed, which can take the following form:

$$H_0'' = \frac{H_0'}{dn}(n_j), \tag{7}$$

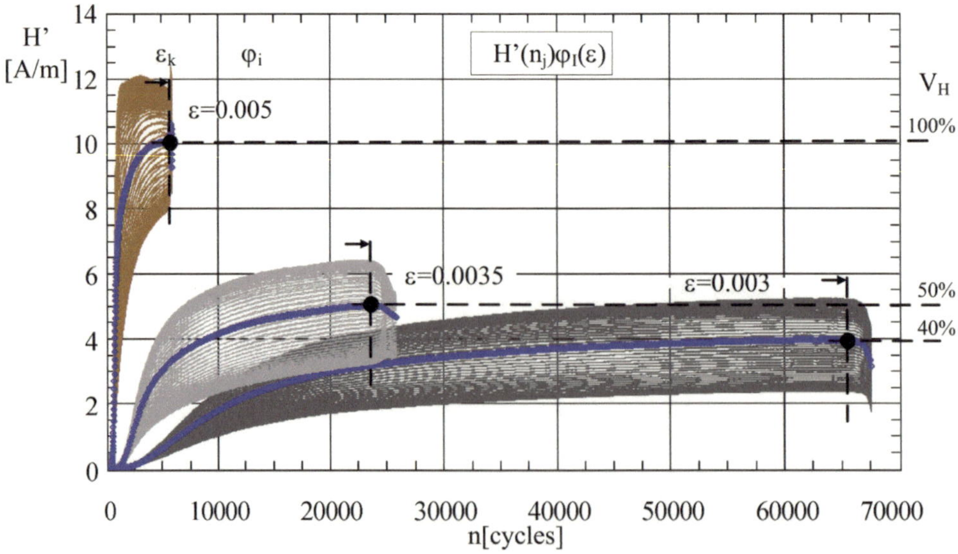

Figure 14. Comparison of H′(ε) with the number of load cycles for three load levels.

http://dx.doi.org/10.5772/intechopen.71408

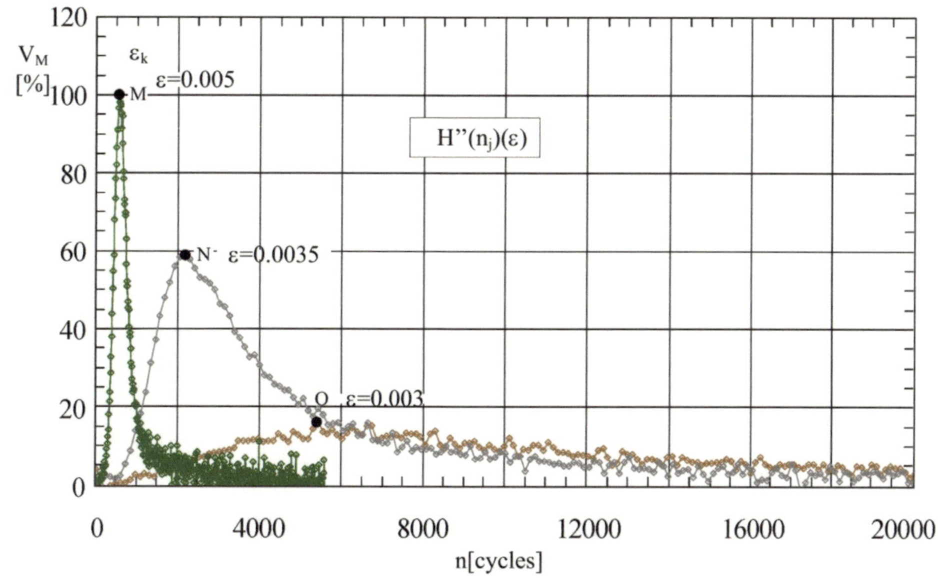

Figure 15. Relative martensite increment speed marked as V_M depending on the level of total strain ε_a.

With reference to the critical value of the total strain level ε_k, it can be stated that

$$V_M = \frac{H_0"(n_j)(\varepsilon_a)}{H_0"(n_j)(\varepsilon_k)} \cdot 100\% \tag{8}$$

In **Figure 15**, $H"_{max}$ points are marked as the maximum speed values of the martensite phase occurrence with regard to the sample with the critical strain level.

4. Application of the thermo-elastic effect: measurement methodology and analysis of results

Temperature was measured using a type E thermocouple made in a classical manner using two wires (CuNi and NiCr) with a diameter of d = 20 μm each. The small mass of wires reduced thermal inertia and enabled the registration of the higher harmonic components of temperature signal ΔT(ξ).

The basic parameters of the thermocouple used in the measurements are presented in **Table 2**.

In the measurements, a universal measurement system consisting of a spectrum analyser HP E1432A and the Keysight VEE software package was used.

Figure 16 presents the cyclic stress-strain curve σ_a-ε_a for the tested material. The cyclic yield limit is $\sigma_a = \sigma_{cpl} = 326$ MPa, and the corresponding value of strain amplitude is $\varepsilon_a(\sigma_{cpl}) = 0.0016$. The curve was determined by connecting the peaks of subsequent hysteresis loops at gradually incrementing load levels. The hysteresis loops are also presented in **Figure 16**. Above

Notation	Materials		Voltage value change to °C depending on temperature			Range of thermocouple work in °C		Thermocouple tolerance	
	+	–	100 [°C]	500 [°C]	1000 [°C]	Continuous work	Temporary work	−40 ÷ 375 [°C]	−40 ÷ 375 [°C]
E	NiCr	CuNi	68	81	–	0 ÷ 800	−40 ÷ 900	±1.5	±0.004*|t|

Table 2. Basic parameters of type E thermocouples.

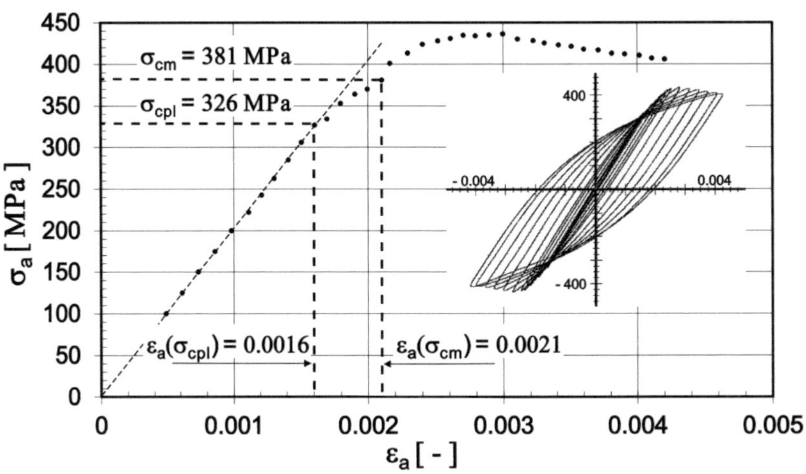

Figure 16. Cyclic stress-strain curve σ_a-ε_a.

the cyclic yield limit, it is possible to observe a characteristic fault on curve σ_a-$\varepsilon_{a,}$ which was identified with the cyclic limit of the martensite transformation $\sigma_{cm} \cong 380$ MPa (corresponding with strain amplitude ε_a $(\sigma_{cm}) = 0.0021$).

Temperature signal $\Delta T(\xi)$ underwent harmonic analysis using the fast Fourier transform (FFT). In the further part of the analysis, the subsequent sinus components of signal $\Delta T(\xi)$ were marked as *1_sin, 2_sin, 3_sin,* etc., whilst the cosine components were marked as *0_cos, 1_cos*, respectively, etc. **Figure 17** presents the change of average temperature value $\Delta T_m(\varepsilon_a)$, which is equivalent to the waveform of the zero cosine component *0_cos*(ε_a).

Temperature value ΔT_m is equal to zero until the strain reached the value $\varepsilon_a = 0.0016$, which corresponds to the cyclic plasticity limit ($\sigma_a = \sigma_{cpl} = 326$ MPa). Later, temperature ΔT_m systematically grows with the increase in strain and, hence, also plastic strain. In **Figure 17**, the initiation moment of the martensite transformation is also marked; however, there is no particular visible trace of this moment in transform $\Delta T_m(\varepsilon_a)$. **Figure 18** presents the set of harmonic components (ignoring *0_cos*) as a load-level function.

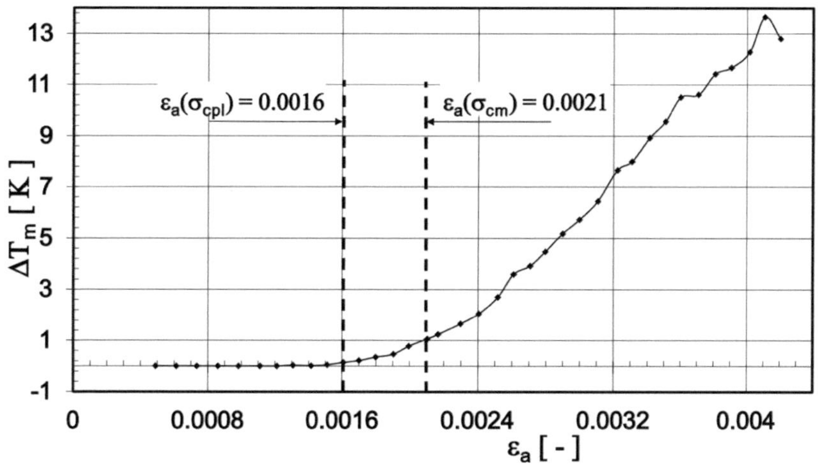

Figure 17. Average temperature dependence on the strain amplitude level $\Delta T_m(\varepsilon_a)$.

In the area in which only the thermo-elastic effect occurs, one can easily observe that the first sinus component (*1_sin*) is linearly dependent on the load level until the cyclic plasticity limit occurs. The influence of higher harmonic components in the elasticity region is negligible.

After crossing the cyclic yield limit, one can observe the deviation of experimental waveform *1_sin* from the linear one, and there are also higher harmonic components. Thus, **Figure 18** confirms the fact that the Thomson model is correct in the elastic range of a paramagnetic material.

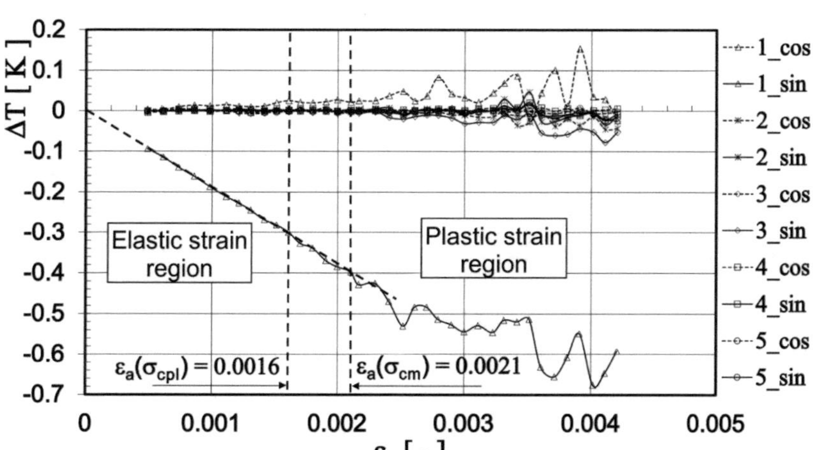

Figure 18. Harmonic components (cosine and sinus) of temperature signal $\Delta T(\xi)$ as a function of strain amplitude ε_a.

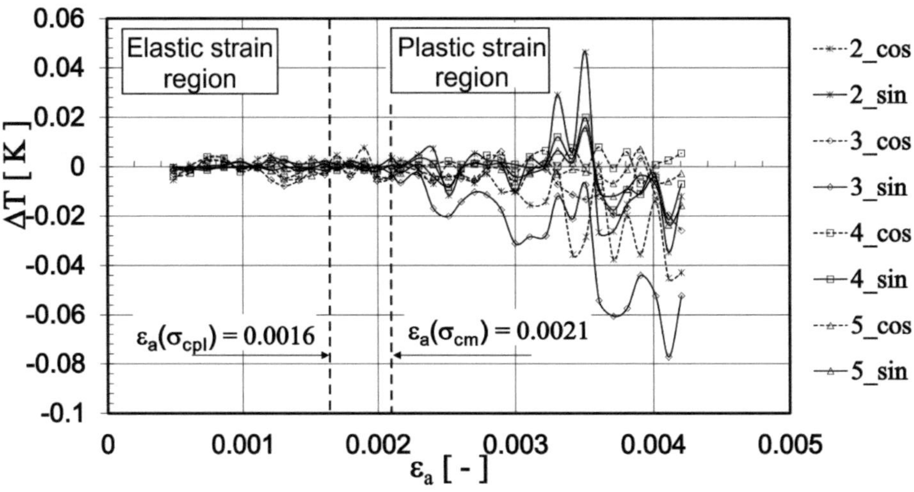

Figure 19. Harmonic components (cosine and sinus) of temperature signal $\Delta T(\xi)$ as a function of strain amplitude ε_a.

Higher harmonic components (except *1_sin*, *0_cos* and *1_cos*) are presented in **Figure 19**. The effect of the so-called saddle effect, analysed earlier in **Figure 1**, is visible here in the form of a clear increment of higher harmonic components, especially the second and fourth expression.

5. Research on the martensite transformation induced by plastic deformation in austenitic foils

The way of investigating the kinetics of the martensite transformation induced by plastic strain in austenite foil of 0.05 mm is presented below. The problem is discussed in detail in [31, 32].

5.1. Preconditions of the measurements of martensite transformation kinetics in austenite foils

The research on materials in the form of a thin foil is a challenging task in terms of conducting measurements. It is necessary to take into consideration the following conditions and requirements:

a. The small size of samples (the cross section of the investigated part in the analysed example was 0.195 mm^2, that is, about 290 less than in the case of the cylindrical sample). This means that force must be applied with an accuracy of 0.1 N. The way of fixing a sample, controlling load and recording measurement signals, both mechanical and magnetic ones, is also of key significance.

b. The necessity to eliminate the negative impact of the magnetic background of a standard magnetic pulsator and to conduct structural research on line (during loading samples) meant that it was necessary to build a miniature testing machine. The small dimensions of the machine and its mass below 1000 g allowed the experiment to also be conducted in the vacuum chamber of a scanning microscope.

c. The need to measure magnetic field strength so as to detect the initiation and increment of ferromagnetic martensite α'. Small increments ΔH required the elimination of field disturbances resulting from both the flow of electric current through the extensometer and the subcomponents of the testing machine made of a ferromagnetic material.

d. Creating dumbbell-type samples of foil required the construction of a special blanking tool which minimised plastic strains on sample edges and, hence, also the initiation place of the martensite transformation. Additionally, samples were tempered using a special procedure.

e. The necessity to eliminate the negative impact of the magnetic background of the standard hydraulic pulsator meant that it was necessary to construct a miniature testing machine.

5.2. Research object, measurement setup, methodology and experiment results

The material was delivered in the form of foil, 0.05 mm thick, in the state marked by the manufacturer as 1/8 ÷ 1/4 of hardness. The foil was heat treated for the purpose of recrystallisation. During the treatment the sample was heated to oversaturation temperature, that is, about 1050°C, and next it was quickly cooled down in air.

After the process of the mechanical cutting of the sample, it was heat treated one more time at a temperature of 1100°C for about 4 s and then cooled down in air. Due to the small size of the samples, they were heated up to a temperature of 1100°C using the flow of electric current of the previously set, experimentally selected parameters. The heat treatment of the samples was conducted on a test setup specially created for this purpose, which consisted of a voltage source (autotransformer) and measurement devices to control temperature, current and voltage.

The conducted heat treatment allowed the obtaining of a single-phase, paramagnetic material with austenite structure, which was free from residual stress, martensite and most dislocations. The average size of grain after annealing was about 40 μm. For the purpose of removing the created in the process magnetic oxides covering the surface of samples, they were polished electrolytically before the research until they were 80 μm thick. The structure of the material when it was delivered and after heat treatment is presented in **Figure 20A** and **B**.

The shape and dimensions of the sample are presented in **Figure 21**.

The main subcomponent was the miniature testing machine (micro-tensile testing machine) presented in **Figure 22**.

Loads were applied using a system with a precise lead screw, worm gear and numerically controlled stepper motors. The drive allowed grips to be shifted in a speed range between 0.095 mm/min and 0.0005 mm/min. The other system ensured transverse movement allowing a sensor to be relocated. This measurement system enabled the recording of mechanical signals (force S(t), displacement Δl(t), strain ε (t)) and changes in magnetic field strength ΔH(t) [23, 33]. Signal processing was conducted using the VEE measurement package and MATLAB.

The determination of the quantitative content of a martensite phase in a material is important, for example, in selecting the technological parameters of plastic forming or welding processes.

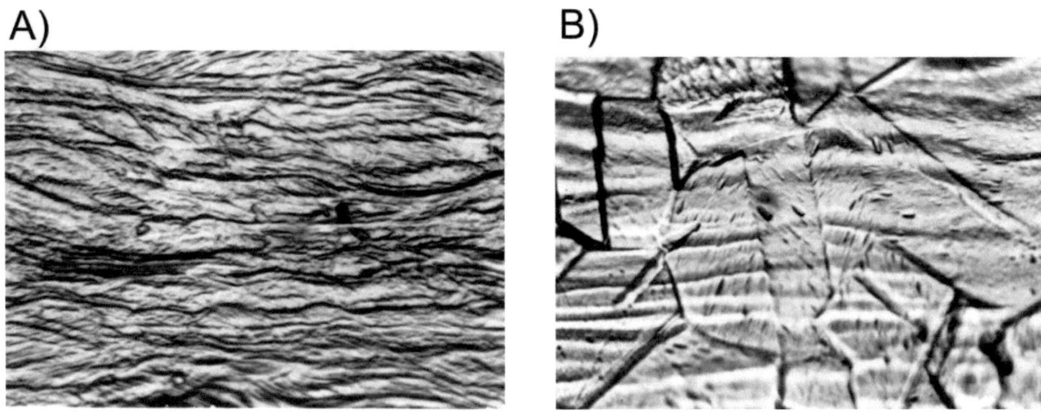

Figure 20. Structure of AISI 304 steel (1000× magnification): before heat treatment (A) and after heat treatment (B) [31].

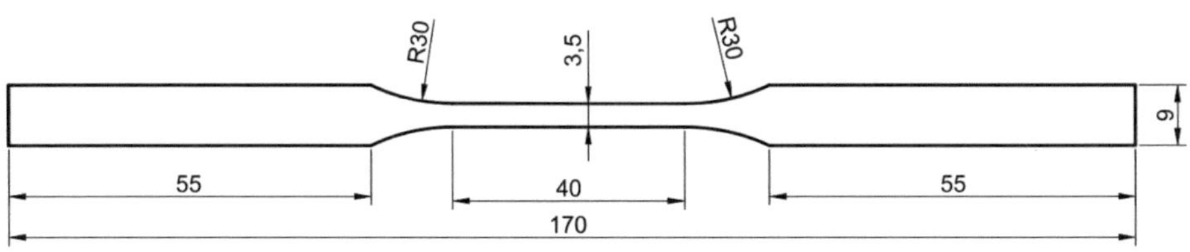

Figure 21. Sample shape and dimensions [31].

There are methods that allow the content of the martensite phase to be determined in laboratory and industrial conditions. Destructive tests encompass:

- Optical microscopy, for example, using phase contrast
- Electron microscopy, using a scanning electron microscope (SEM)
- Roentgen methods

One of the most effective methods of determining the amount of residual austenite in steels is the roentgen method. It examines the mutual intensity ratio of X-ray reflections, in particular phases, which were diffracted on the crystalline network of the examined steel. X-rays are diffracted on crystalline networks at various Bragg angles. This allows identification of these phases in steel and calculation of their percentage content, including also residual austenite. This is why the research was conducted using both the magnetic method and a diffractometer. The measurements were made using the test setup based on the micro-tensile testing machine. During the work, it turned out that the sensitivity of magnetic sensors was high enough to also register the movement of the mechanism in the miniature testing machine. As a result, a mobile measurement beam made of aluminium and a sample with a dynamometer system were located at a significant distance from the ferromagnetic parts of the micro-tensile testing machine. An important source of interference turned out to be the tensometric bridge of the dynamometer, the problem of which was solved by using lower voltage than before. As

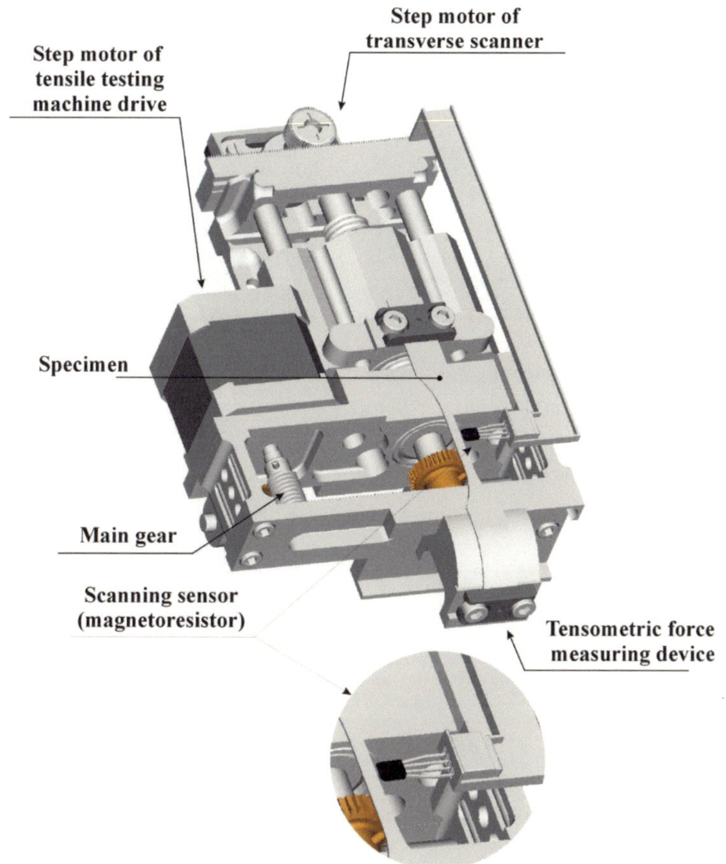

Figure 22. Construction of the micro-tensile testing machine with a description of its main components [31].

a result of these actions, at least a 10-fold lower influence of the interfering magnetic field in the background was obtained, which enabled the high resolution of magnetic field increment measurement 0.01 A/m.

The tested foil was placed in non-magnetic grips, and the controlled process plastic strain was started. **Figure 23** shows the pictures of austenite foil fixed in the grips of the measurement device in which it was broken.

Figure 24 presents the tension curve (A) and its magnetic response (B) for the tested austenite foil. The increase in magnetic signal was the consequence of the appearance and increase in martensite α', a ferromagnetic material in the paramagnetic austenite matrix.

The tests on foil were also conducted using an Xstress 3000 diffractometer. The sample was exposed to radiation at various values of strain. The obtained results are presented in **Figure 25**. The Y-axis is the intensity of X-ray captured by the matrix CCD on which there is a beryllium screen transmitting only a particular wavelength of radiation. The X-axis is the CCD matrix resolution.

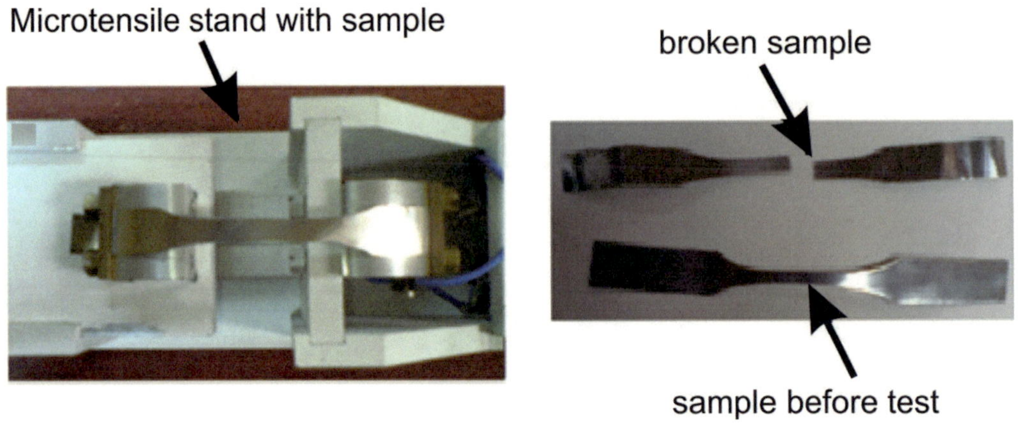

Figure 23. Thin austenite foils: sample fixed in the grips of the micro-tensile testing machine, sample before measurement and after breaking [31].

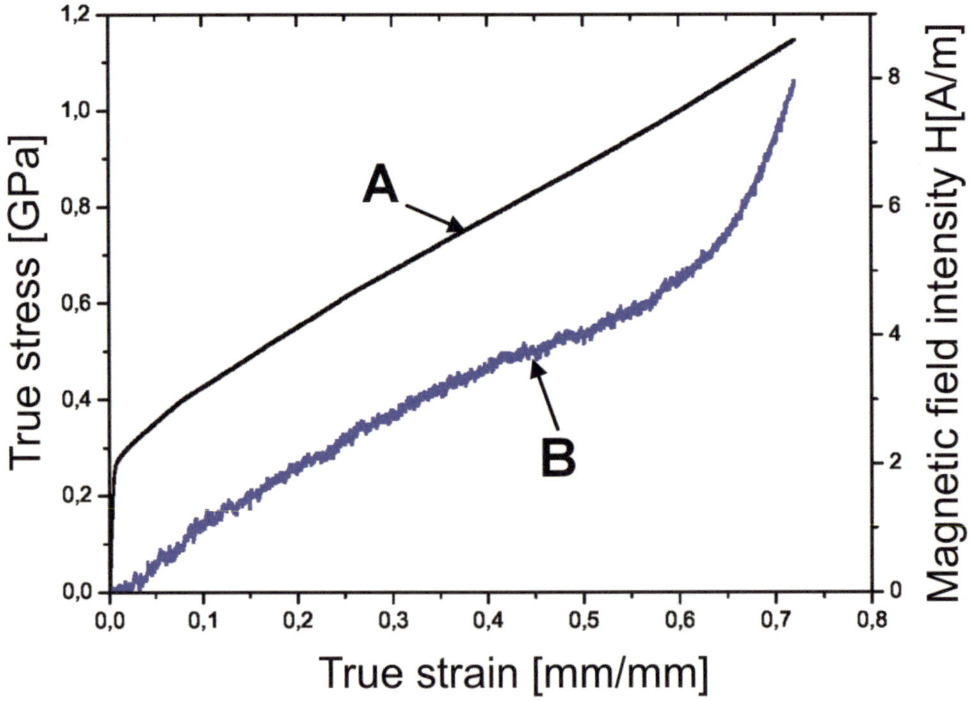

Figure 24. Results of plastic strain of the austenite foil sample: quasi-static stress-strain curve of the material and magnetic response to plastic strain [18].

As was demonstrated before, the changes of the magnetic properties of a material resulting from its strain are strictly related to the kinetics of the examined martensite transformation induced by plastic strain [31, 32]. It was shown in earlier works [32] that the first martensite nuclei appear after exceeding $\varepsilon = 0.035$, which is recorded by both magnetic field sensors and an Xstress 3000 diffractometer.

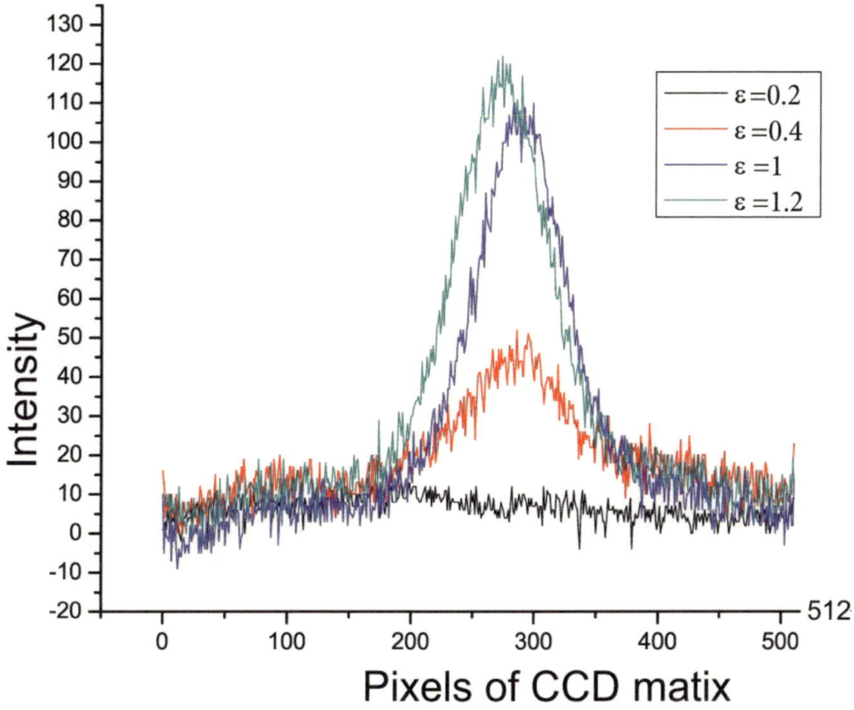

Figure 25. Images obtained from the CCD matrix: spectrum intensity change for selected strain levels [18].

Foil samples were loaded at the speed of 0.035 mm/min. In the introductory phase of each experiment, the goal was to achieve initial stress so as to eliminate the corrugation of the sample. Next, the main part of the experiment was conducted until the sample was broken. Mechanical magnetic signals were recorded all the time.

Figure 26 presents the experimental dependencies of force S(t), strain ε(t) and magnetic field strength H(t). It should be emphasised that the strain signal could only be measured to the value of $\varepsilon \approx 0.0012$, and after reaching this value the sensor was destructed.

The analysis of signals in the initial phase of strain is clearer if waveforms are presented, for example, until time t = 16 minutes, which is presented in **Figure 27**. It is then possible to distinguish a few phases of the process. The first stage takes place in the time range $0 \leq t \leq 3.7$ minutes and is related to flexing the foil. In the range $3.7 \leq t \leq 7.8$ minutes, the elastic load of material takes place. The force signal changes linearly (except for a short transient period) and is accompanied by a strain change that is also linear. The value of signal H(t) is equal to zero if small fluctuations related to the measurement method are neglected.

At time $t \approx 7.8$, one can observe a bend in the force course and a change in the inclination angle of course ε(t), which confirms the occurrence of yield limit. It is also the beginning of the clear and constant increment of signal H(t), which, in turn, confirms the appearance of the ferromagnetic phase (martensite α'). It is difficult to simultaneously separate in the chart the yield limit and value σ_m which marks the initiation of the phase transformation. The further growth

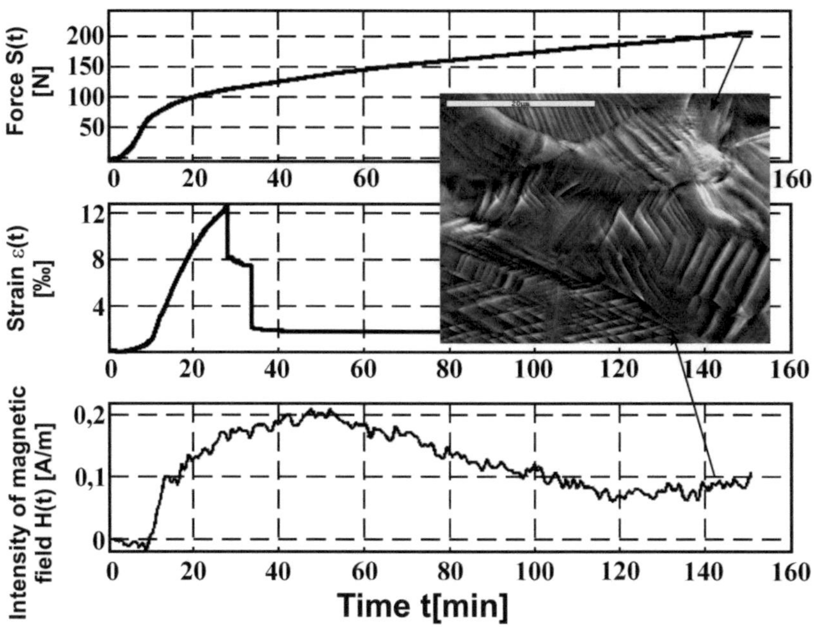

Figure 26. Characteristics F = f(t), ε = f(t) and H = f(t). AISI 304 steel, speed 0.035 mm/min. Additionally, the image material structure in the final part of strain (scanning microscope) [31].

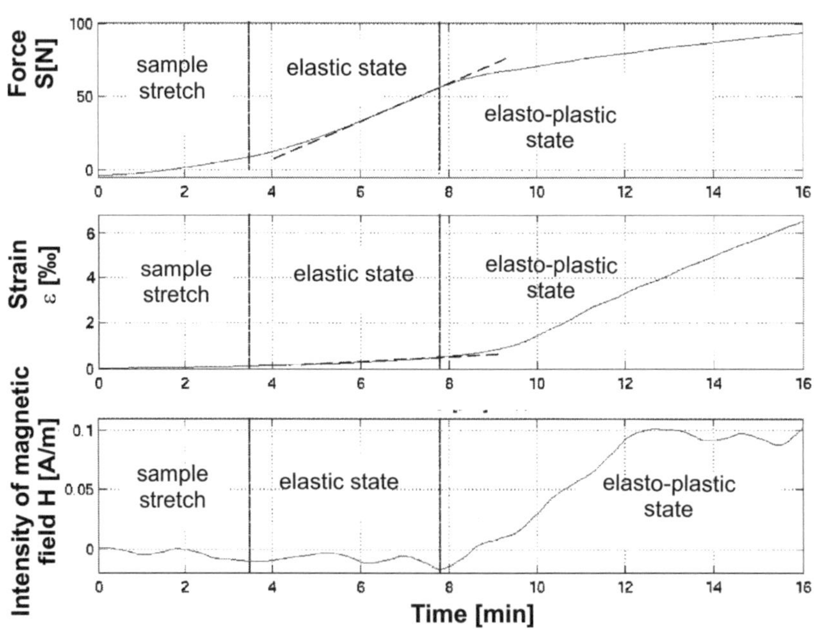

Figure 27. Characteristics F = f(t), ε = f(t) and H = f(t). AISI 304 steel, speed 0.035 mm/min (magnification) [31].

of strain in the plastic area results in the increment of H(t) until t ≈ 50 minutes and then there is a fall. This means the increment of martensite α' on the one hand and on the other hand blocking the movement of domains n.

http://dx.doi.org/10.5772/intechopen.71408

Research is currently being conducted on static strain with the simultaneous observation of changes in the structure. **Figure 26** shows a sample image of the material structure in the final phase of foil strain.

6. Conclusions and final remarks

The methodology of the research on the kinetics of the martensite transformation induced by plastic strain has been developed. Two cross effects (the Villari and Kelvin effects) were used for this purpose. This allowed transformations during the fatigue process to be followed and visualised on line, without the necessity to use, for example, roentgenographic or microscopic methods. The main conclusions are presented below and are divided according to the type of samples and the used cross effects:

1. Massive samples and magneto-mechanical effect:
   - **a.** The measurements were conducted with the passive, contactless method using the quasi-pointwise sensors of a magnetic field (magnetoresistors). These sensors are very sensitive wideband devices (they can operate between a DC signal and a AC signal at the frequency of a few MHz).
   - **b.** After exceeding the cyclical yield limit, the martensite transformation was initiated in a paramagnetic austenite sample, and the increment of ferromagnetic martensite α' started, which was registered by a measurement head with magnetoresistors.
   - **c.** The hydraulic pulsator and the sample form a complex magnetic system. In the measurement methodology, the assumption that the range of field changes along the sample near the zero line is linear, regardless of this obstacle, was of key significance.
   - **d.** The notion of the zero magnetic line of the sample was introduced. Magnetic values were defined: H'_0, average increment of a magnetic field per unit of length of the measurement part of the sample as the magnetic measure of martensite content in the sample, and H''_0, martensite increment speed, respectively. Both values depend on the level of the mechanical load of the material and the number of cycles. Values H'_0 and H''_0 should be connected with the amount of martensite using calibration with other methods (e.g. roentgenographic or microscopic).
   - **e.** High-resolution phase maps allowing the use of the image processing methods of such well-known techniques as machinery visioning (MV) or digital image correlation (DIC) were used in the visualisation of magnetic field changes.
2. Massive samples and thermo-elastic effect:
   - **a.** The thermo-elastic effect can be used to determine the cyclic yield limit σ_{cpl}. For stresses $\sigma_a < \sigma_{cpl}$, the harmonic spectrum of temperature signal $\Delta T(\xi)$ only includes component 1_sin. For stresses $\sigma_a > \sigma_{cpl}$, higher harmonic components become visible.
   - **b.** The moment of the initiation of the martensite transformation is manifested by the increment of the selected values of higher harmonic components (especially the second and fourth ones).

c. The measurement method, the used apparatus, the software and the thermocouples with small thermal inertia showed that the thermomechanical effect was fully usable in the research on the fatigue process and the martensite transformation.

3. Foil samples and magneto-mechanical effect:

a. The martensite transformation induced by plastic strain took place in a foil 0.05 mm thick that was made of metastable austenite steel AISI 304. It was confirmed by the course of martensite signal H(t), in which the increment was observed as a result of the occurrence of the ferromagnetic phase (martensite α') and the recording of the Villari effect.

b. An original, numerically controlled test machine of 800 g was developed and made with a measurement system with mechanical and magnetic values. It proved to be very useful in the research on the phase transformation in foil samples. The device enables material strain, recording mechanical and cross signals with simultaneous observation of structural changes for the purpose of conducting the quantitative assessment of the martensite phase increment.

c. The martensite transformation in such thin austenite tapes is manifested by the occurrence of a very weak magnetic field (maximum 0.2 A/m), which is on the border of the maximum measurement resolution of magnetoresistors. A new more precise method of measuring a magnetic field that takes into account SQUID detectors should be considered. Additionally, it was experimentally proven that the sensitivity of magnetoresistant sensors decreases with an increase in temperature.

4. Further research

Further research will focus on the calibration of the results obtained using both cross effects. Microscopic and roentgenographic methods will be employed. Such a calibration is required for each type of metastable austenite steel in which the martensite transformation induced by plastic strain takes place.

Acknowledgements

The research was funded by project 0401/0029/17.

Author details

Jerzy Kaleta*, Przemysław Wiewiórski and Wojciech Wiśniewski

*Address all correspondence to: jerzy.kaleta@pwr.edu.pl

Department of Mechanics, Materials Science and Engineering, Wroclaw University of Science and Technology, Wrocław, Poland

References

[1] Ziętek G, Mróz Z. Thermomechanical model for austenitic steel with martensitic transformation induced by temperature and stress variation. In: Book of Abstracts of the 38th Solid Mechanics Conference, SolMech. Warsaw: Institute of Fundamental Technological Research (IPPT) of the Polish Academy of Sciences; 2012. p. 114-115

[2] Ziętek G, Mróz Z. On the hardening rule for austenitic steels accounting for the strain induced martensitic transformation. The International Journal of Structural Changes in Solids. 2011;**3**(3):21-34

[3] Ziętek G, Mróz Z. Description of cyclic hardening of material with plasticity induced martensitic transformation. In: Kotulski Z, Kowalczyk P, Sosnowski W, editors. Selected Topics of Contemporary Solid Mechanics: Proceedings of the 36th Solid Mechanics Conference. Warszawa: Instytut Podstawowych Problemów Techniki PAN; 2008. p. 430-431

[4] Talonen J, Aspegren P, Hänninen H. Comparison of different methods for measuring strain induced α-martensite content in austenitic steels. Materials Science and Technology. 2004;**20**(12):1506-1512

[5] Nagy E, Mertinger V, Tranta F, Sólyom J. Deformation induced martensitic transformation in stainless steels. Materials Science and Engineering: A. 2004;**378**(1):308-313

[6] Amitava M, De P, Bhattacharya D Sr, Srivastava P, Jiles D. Ferromagnetic properties of deformation-induced martensite transformation in AISI 304 stainless steel. Metallurgical and Materials Transactions A. 2004;**35**(2):599-605

[7] Kaleta J, Zietek G. Cyclic plastic deformation-induced martensitic transformation in austenitic steels. Advances in Fracture Research, ICF 9. 1997;**1**:275-281

[8] Hahnenberger F, Smaga M, Eifler D. Microstructural investigation of the fatigue behavior and phase transformation in metastable austenitic steels at ambient and lower temperatures. International Journal of Fatigue. 2014;**69**:36-48

[9] Tavares S, Fruchart D, Miraglia S. A magnetic study of the reversion of martensite α' in a 304 stainless steel. Journal of Alloys and Compounds. 2000;**307**(1):311-317

[10] Lang M, Johnson J, Schreiber J, Dobmann G, Bassler HJ, Eifler D, Ehrlich R, Gampe U. Cyclic deformation behaviour of AISI 321 austenitic steel and its characterization by means of HTC-SQUID. Nuclear Engineering and Design. 2000;**198**(1):185-191

[11] Shin HC, Ha TK, Chang YW. Kinetics of deformation induced martensitic transformation in a 304 stainless steel. Scripta Materialia. 2001;**45**(7):823-829

[12] Kaleta J, Lewandowski D, Wiśniewski W. Kinematics of cross efects in cyclic load induced martensitic transformation in massive specimens. In: Scientific Letters of Kielce University of Technology, Kielce. Vol. 73. 2001. p. 391-400

[13] Kaleta J, Tumański S, Żebracki J. Magnetoresistors as a tool for investigation of mechanical properties of ferromagnetic materials. Journal of Magnetism and Magnetic Materials. 1996;**160**:165-166

[14] Kaleta J, Susz S. Magnetovision system for examination of biaxial stress state under cyclic load. In: Experimental Mechanics of Solids - 17th Symposium, Jachranka, Poland, 1996. p. 290-295 (in Polish)

[15] Thomson W. On the dynamical theory of heat. Transactions of the Royal Society of Edinburgh. 1853;**20**:261-283

[16] Kaleta J, Moczko P, Wiśniewski W. Harmonic analysis of temperature signal in thermoelastic effect for austenite under cyclic load. In: Experimental Mechanics of Solids - 19th Symposium, Jachranka, Poland, 1996. Vol. 2000. p. 296-301 (in Polish)

[17] Kaleta J, Ziętek G. Cyclic plastic deformation-induced martensitic transformation in austenitic steels. In: Proceedings of 9th International Conference on Fracture ICF-9, Pergamon, Sydney. Vol. 1. April 1-5, 1997. p. 275-281

[18] Kaleta J, Wiśniewski W, Wiewiórski PK. Magnetic method of testing martensitic transformation induced by plastic deformation in solid specimens of AISI 304. In: Symposium on Fatigue and Mechanics of Materials and Structure, Bydgoszcz-Pieczyska. 2008. p. 139-145 (in Polish)

[19] Kaleta J, Wiewiórski PK, Wiśniewski W. Methods of testing martensitic transformation induced by plastic deformation in solid specimens. In: Experimental Mechanics of Solids - 23th Symposium, Jachranka, Poland, 2008 (in Polish)

[20] Kaleta J, Żebracki J. Application of the Villari effect in fatigue examination of nickel. Fatigue and Fracture of Engineering Materials and Structures. 1996;**19**(12):1435-1443

[21] Kaleta J, Wiewiórski P. Magnetic field distribution detecting and computing methods for experimental mechanics. Engineering Transactions. 2010;**58**(3/4):97-118

[22] Kaleta J, Wiewiórski P. Magnetovisual method for monitoring thermal demagnetization of permanent magnets used in magnetostrictive actuators. Journal of Rare Earths. 2014;**32**(3):236-241

[23] Tumański S, Stabrowski M. Magnetovision system, new method of investigating steel sheets. Journal of Magnetism and Magnetic Materials. 1996;**160**:165-166

[24] Philips Semiconductor "Magnetoresitive Sensors for Magnetic Field Measurement", Application Manual SC17 Rev. 2, Sep 6, 2000

[25] Gasparics A. Electromagnetic Nondestructive Material Evaluation Based on Fluxset Sensor, [PHD thesis]. Budapest University of Technology and Economics; 2005

[26] Yang CH, Tsai MC. Measurement of magnetic fields by B-spline method. Journal of Magnetism and Magnetic Materials. 2006;**304**:460-463

[27] Tan S, Ma YP, Thomas IM, Wikswo JP Jr. High resolution SQUID imaging of current and magnetisation distributions. IEEE Transactions on Applied Superconductivity. 1993;**3**(1):1945-1948

[28] Lo CCH, Paulsen JA, Jiles DC. A magnetic imaging system for evaluation of material conditions using magnetoresistive devices. IEEE Transactions on Magnetics. 2003;**39**(5):3453-3455

[29] Infante F, Perdu P, Lewis D. Magnetic microscopy for ground plane current detection: A fast and reliable technique for current leakage localization by means of magnetic simulations. Microelectronics Reliability. 2010;**50**:1700-1705

[30] Dodd RK. A new approach to the visualization of tensor fields. Graphical Models and Image Processing. 1998;**60**:286-303

[31] Dubov A, Kolokolnikov S. The metal magnetic memory method application for online monitoring of damage development in steel pipes and welded joints specimens. Weld World. 2013;**57**:s123-s136

[32] Fassa B, Kaleta J, Wiewiórski PK. Plastic strain induced martensitic transformation in foils with metastable austenite. In: Scientific Letters of Kielce University of Technology 2001. p. 343-350 (in Polish)

[33] Fassa B, Kaleta J. Identification of kinetics martensitic transformation induced by plastic strain in austenitic foils. In: Scientific Letters of Kielce University of Technology. Vol. 4, 2007. p. 27-29 (in Polish)

Chapter 4

Effect of Sigma Phase on Fracture Behavior of Steels and Weld Joints of Components in Power Industry Working at Supercritical Conditions

Zdeněk Kuboň, Šárka Stejskalová and
Ladislav Kander

Additional information is available at the end of the chapter

http://dx.doi.org/10.5772/intechopen.71569

Abstract

This chapter deals with the change of the structure and mechanical properties of the austenitic steels grades TP347HFG, Super 304H and HR3C as well as their welded joints after the exposure at temperatures corresponding to the ultra-super critical (USC) and advanced ultra-super critical (A-USC) parameters. Several tube bend radii were investigated in order to understand relation between cold work plastic deformation, mechanical properties and the structural changes with special attention to the generation of sigma phase during long-term high temperature exposure. The effect of post bend solution annealing work on the material properties and structural changes was studied, too, using small punch test (SPT) and miniaturized tensile tests taken both from straight part and extrados of bends under investigation. Creep properties of either base metal either welded joints were tested and evaluated using standard creep specimens at temperatures 650, 700 and 750°C. The obtained results confirmed that the mechanical properties and the structure were significantly influenced even after relatively short-time exposure at elevated temperature. The results of experiments were compared with other results obtained from analyses of tubes after 100,000 h of exposure in a USC block where extensive precipitation of sigma phase was also identified.

Keywords: σ-phase, austenitic steels, tube bends, weld joints, creep properties

1. Introduction

The utilization of ultra-super critical (USC) boilers working with advanced steam parameters (typically 600/620°C and 28.5 MPa) has led to the development of new materials with improved long-term creep properties. In order to meet the demanding conditions, new grades of low-alloy steels for waterwalls, modified chromium steels for headers and pipelines as well as austenitic steels and nickel superalloys for the final stages of superheaters and reheaters (RHs) have been developed. The investigation of the functional properties and further development of groups of materials is thus the key factor in the design and construction of new power units.

The effort to improve creep resistance of austenitic steels based on 18Cr-8Ni series, originally developed as corrosion resistant materials, was initiated in 1970s. Similarly, the improvement of creep resistance of 25Cr-20Ni series for demanding corrosive conditions has started in 1980s. **Figure 1** shows the development of austenitic creep-resistant steels for fossil fired power plant applications [1].

Austenitic heat-resistant steels for boiler applications of 18Cr-8Ni series are used in the under-stabilized condition, the content of strong carbide forming elements (Ti, Nb) is decreased and $M_{23}C_6$ carbide contributes to the precipitation strengthening of the steel. Thanks to the working conditions at high temperatures, austenitic creep-resistant steels do not need to be resistant to the intergranular corrosion caused by precipitation of $M_{23}C_6$ carbides on grain boundaries and high temperature also enable fast healing effect due to fast diffusion of chromium to the chromium depleted zone around carbide on the grain boundary.

Figure 2 shows the increase of creep rupture strength of heat-resistant boiler steels at 600°C after 100,000 h for materials developed during the twentieth century [2].

2. Advanced austenitic steels for USC boilers and their material characteristics

The austenitic steels of AISI 304(H) and AISI 316(H) type used in boilers, where the temperature of the superheater tubes can exceed 650°C, suffer from significant steam-side oxide

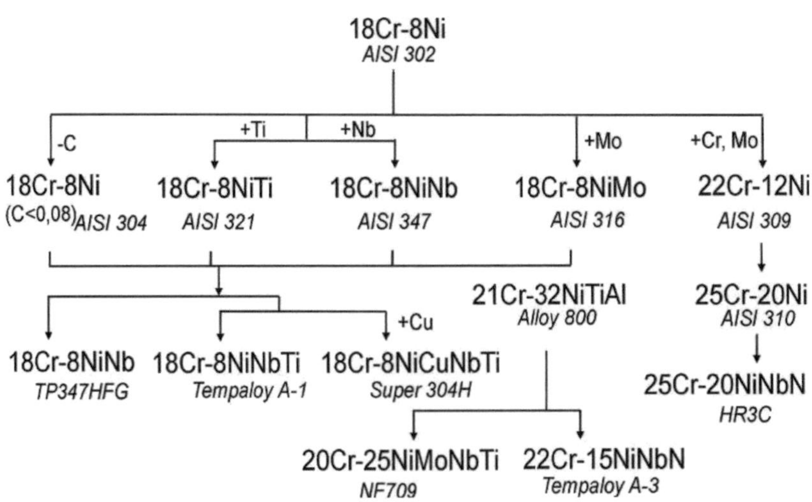

Figure 1. Development of austenitic creep-resistant steels for power plants [1].

http://dx.doi.org/10.5772/intechopen.71569

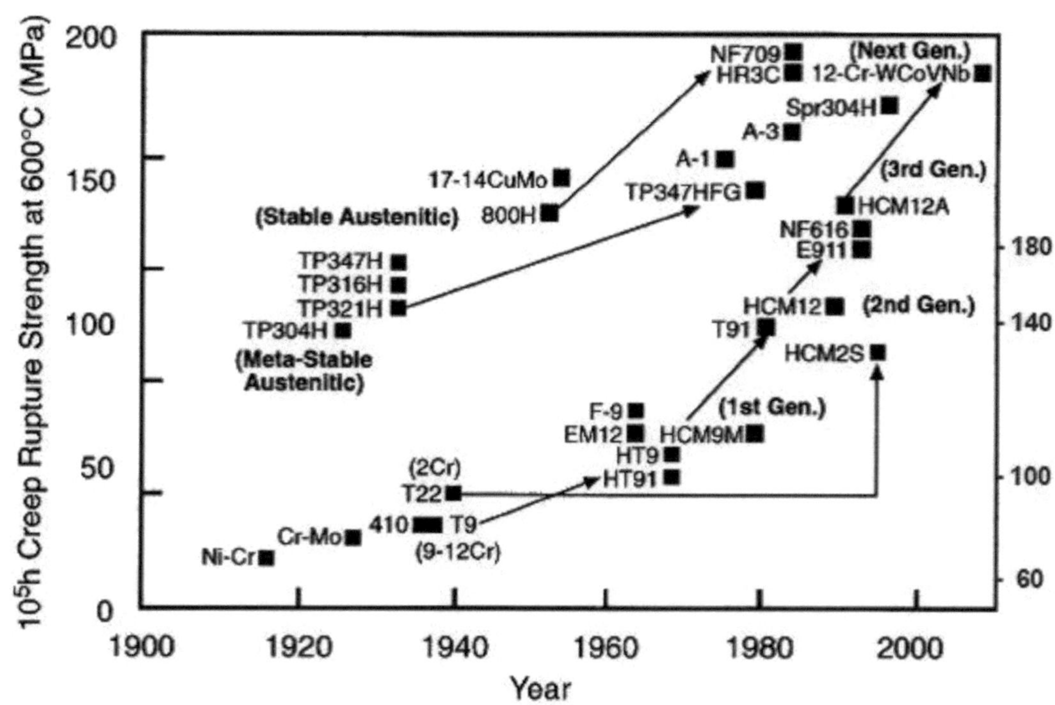

Figure 2. Increase of 100,000 h creep rupture strength of types and grades of creep-resistant steels at 600°C [2].

spallation resulting in overheating and subsequent creep fracture of the tubes [3]. This spallation occurs due to different thermal expansion of the steel and the oxides on the steam-side surface of the tube. The compactness of the oxide layer in austenitic steels then causes spallation of the almost entire protective oxide layer during sudden temperature changes.

Efforts to increase the steam oxidation resistance as well as creep strength at temperatures above 650°C resulted in the development and successful use of advanced austenitic heat-resistant steels for USC boilers known as Super 304H (1.4907, X10CrNiCuNb 18-9-3), HR3C (1.4952, X6CrNiNbN 25-20) and TP347HFG (1.4908, X8CrNi 19-11). The chemical composition and mechanical properties are shown in **Tables 1** and **2**, respectively.

Chemical composition of steel Super 304H has been changed in comparison to the grade AISI 304H by the addition of 3% Cu and alloying by niobium and nitrogen. Copper forms particles of ε-phase in the matrix, that is, small spherical precipitates that significantly strengthen the matrix. Minimal changes were made to TP347HFG compared to the former grade TP347H, the niobium to carbon ratio was increased and special heat treatment that guaranties a fine-grained structure was developed. Steel HR3C has balanced nitrogen and niobium contents, which should increase creep resistance and reduce the susceptibility to the formation of undesirable hard and brittle σ-phase [4]. The change of chemical composition, namely addition of niobium increased precipitation strengthening of these steel grades and copper in Super 304H steel further increased its creep rupture strength. All of them (TP347H, Super 304H and HR3C) have been successfully used as superheater/reheater tubes in USC power plants all over the world.

Grade	C	Si	Mn	P	S	Cr	Ni	B	Cu	Nb	Al	N
Super 304H	0.07 0.13	max. 0.30	max. 1.00	max. 0.040	max. 0.010	17.0 19.0	7.5 10.5	0.001 0.010	2.50 3.50	0.30 0.60	0.003 0.030	0.05 0.12
HR3C	max. 0.10	max. 1.50	max. 2.00	max. 0.030	max. 0.030	23.0– 27.0	17.0 23.0	—	—	0.20 0.60	—	0.15 0.35
TP347HFG	0.06– 0.10	max. 0.75	max. 2.00	max. 0.040	max. 0.030	17.0 20.0	9.0 13.0	—	—	8 × C	—	—

Table 1. Chemical composition of advanced austenitic steels (wt.%).

Grade	$R_{p0.2}$	R_m	$R_{p0.2}$ (650°C)	A	KV (+20°C)	R_{uT}/650°C/10^5 h
	(MPa)	(MPa)	(MPa)	(%)	(J)	(MPa)
Super 304H	≥235	590–850	135	≥35	85	116
HR3C	≥295	655–900	180	≥30	85	114
TP347HFG	≥205	550–750	126	≥35	85	100

Table 2. Mechanical properties of advanced austenitic steel grades.

3. Precipitation of sigma phase

The σ-phase is an intermetallic phase that forms in more than 50 transition alloys. The technically most important is σ-phase precipitation in Fe-Cr system. The σ-phase has a tetragonal crystal lattice and precipitates between 600 and 1000°C [5–7]. The chromium contents in the σ-phase containing alloys is usually in range from 25 to 76 wt.%. In alloys with lower chromium content can sigma phase contain other alloying elements as well. In the Cr-Ni austenitic steels, sigma phase formation is encouraged by increasing Cr content above 17–20 wt.%, and is discouraged by an increase in Ni content. Since the sigma phase is Cr-rich phase, the diffusion of Cr is usually regarded as one of the controlling factors for its precipitation. Besides chromium and carbon contents, many other factors affect the formation of σ-phase. Diffusion of another substitutional element like Mo could also be important [8]. In general, all ferrite-forming elements like Cr, Nb, Ti, Mo or W promote precipitation of σ-phase. Silicon also promotes and accelerates σ-phase formation. Due to differences in diffusion rate of alloying elements, the formation of σ-phase in austenite is about 100 times slower than in ferrite. Consequently, the presence of δ-ferrite in austenitic alloys accelerates σ-phase precipitation [9]. The δ-ferrite (BCC), compared to the γ-austenite (FCC), is a chromium-rich region where the diffusion of chromium and other ferrite-forming elements is faster. Hence, δ-ferrite is naturally a beneficial site for the precipitation of the σ-phase [10]. Therefore, in steels containing some ferrite or in which ferrite forms temporarily during aging, will σ-phase be formed via the ferrite.

The σ phase forming tendency of an alloy can be predicted according to the electron vacancy number originally developed by Woodyatt et al. [11] in the form

$$N_V = 0.66Ni + 1.71Co + 2.66Fe + 4.66(Cr + Mo + W) + 5.66V + 6.66Zr + 10.66Nb[at.] \quad (1)$$

When the calculated value of electron vacancy number is greater than 2.52, σ-phase should form.

Since the σ-phase presence is closely related to the $M_{23}C_6$, there is a preposition that $M_{23}C_6$ acts as the precursor to the σ-phase [12–14]. Pre-existence of carbides in the steel can be an intensifying factor for σ-phase formation, as the formed carbides have high chromium content and can act as the source of chromium. The formation from M_7C_3 carbides is faster than from $M_{23}C_6$, probably due to instability of M_7C_3 carbide [15].

Other authors [16] stated that in situ transformation of $M_{23}C_6$ to the σ-phase is improbable, although they tend to be in contact. A study on the precipitation sequence in 316L steel at temperatures up to 500°C has found that the formation of the σ-phase precedes that of $M_{23}C_6$. This work has pointed out that the precursors to the σ-phase in this temperature range are in fact precipitates forming on dislocations and grain boundaries, which are not possible to identify even by using high resolution TEM (HR-TEM) because of their extremely small sizes [17].

In stainless steels, classification of the morphology of the σ-phase can be divided into four types [4]:

1. Grain boundary precipitation,
2. Triple point precipitation,
3. Corner precipitation and
4. Cellular precipitation.

3.1. δ-Ferrite/austenite grain boundary precipitation

The σ-phase easily precipitates at the δ/γ phase boundary, which is a region with high concentration of chromium. At the same time, it is a high interface energy site and a beneficial site for the heterogeneous nucleation thank to high concentration of lattice defects. When the σ-phase nucleates at the δ/γ interphase boundary, some defects disappear, which releases the free energy of the system. Consequently, the activation energy barrier to form a coherent interface is reduced. This case is typical for austenitic stainless steels.

3.2. Triple point precipitation

The σ-phase precipitates also at the triple point of the δ-ferrite boundary. According to Barcik [9], σ-phase precipitates first on triple points and then on grain faces. After long-term aging at high temperature, it also forms on the coherent twin boundaries and intragranular inclusions.

3.3. Corner precipitation

The corner precipitation of the σ-phase means that it precipitates directly at the corner δ-ferrite particles because the δ-ferrite has a high Cr content phase, and σ-phase prefers to nucleate and precipitate at that point. When the σ-phase precipitates at the corner δ-ferrite, it consumes the Cr from the δ-ferrite particles.

3.4. Cellular precipitation

Cellular precipitation means that the σ-phase and secondary austenite ($\sigma + \gamma_2$) precipitate as laminar precipitation in the δ-ferrite particles. This reaction is called the eutectoid decomposition of $\delta \rightarrow \sigma + \gamma_2$. When the eutectoid decomposition of $\delta \rightarrow \sigma + \gamma_2$ is finished, the σ-phase consumes the Cr, Mo and Si from the δ-ferrite particles.

The direct precipitation of σ-phase in austenite is in general very slow and takes thousands of hours due to:

- very low carbon and nitrogen solubility in σ-phase, which means that the both elements have to form carbide and nitride before σ-phase precipitation,
- very slow diffusion rate of substitutional elements in austenite and
- incoherent boundary between σ-phase and austenite, which makes σ-phase nucleation difficult.

Figure 3 shows the precipitation mechanism of σ-phase in AISI 316 L stainless steel at different aging temperatures. Without heating (t_0), the δ-ferrite precipitates at the δ/γ interphase boundary and γ-phase. When the aging temperature increases to t_1, the σ and γ_2-phases precipitate in δ-ferrite particles. When the aging temperatures increases to t_2, the cellular $\sigma + \gamma_2$ also forms in δ-ferrite particles, and the σ-phase precipitates at the triple points and the δ/γ interphase boundaries. When the aging temperature is t_3, the precipitation of the laminar $\sigma + \gamma_2$ is more pronounced than at the other aging temperatures [18].

When the δ-ferrite to σ-phase transformation occurs, the σ-phase precipitates in the Cr-enriched region of δ-ferrite, directly in δ-ferrite particles. The precipitation of σ-phase is a diffusion-controlled phase transformation in 25% Cr–20% Ni stainless steel, where Cr atoms have an important effect on acceleration of the precipitation rate of σ-phase [19]. The precipitation rate of σ-phase in δ-ferrite is therefore about 100 times higher than in γ-austenite [9, 20]. After σ-phase precipitation, will the content of Cr and Mo in δ-ferrite decrease and simultaneously will Ni content increase. This fact leads to the formation of secondary austenite phases (γ_2) through the δ-ferrite [21]. The plate σ-phase and γ_2-precipitate in δ-ferrite particles were observed in microstructures of AISI 304 stainless steel aged at 595 and 650°C for 31,000 h [22]. The precipitation

http://dx.doi.org/10.5772/intechopen.71569

mechanism is through the eutectoid decomposition of δ-ferrite into σ-phase and γ_2-austenite and is similar at both temperatures, but at 650°C it was more evident (see **Figures 4** and **5**).

In stabilized grades, the formation of σ-phase is faster than in other grades. Minami et al. reported precipitation of σ-phase in most of the grades of austenitic stainless steels: type AISI 304, 316, 321 and 347, but it forms after different times (in 347 and 321 already after 1000 h at 700°C but in 304, 316 and Tempaloy-A1, σ-phase was found in significant quantities only after 10,000 h, see **Figure 5** (the percentage of σ-phase equals to the area etched by KOH) [23].

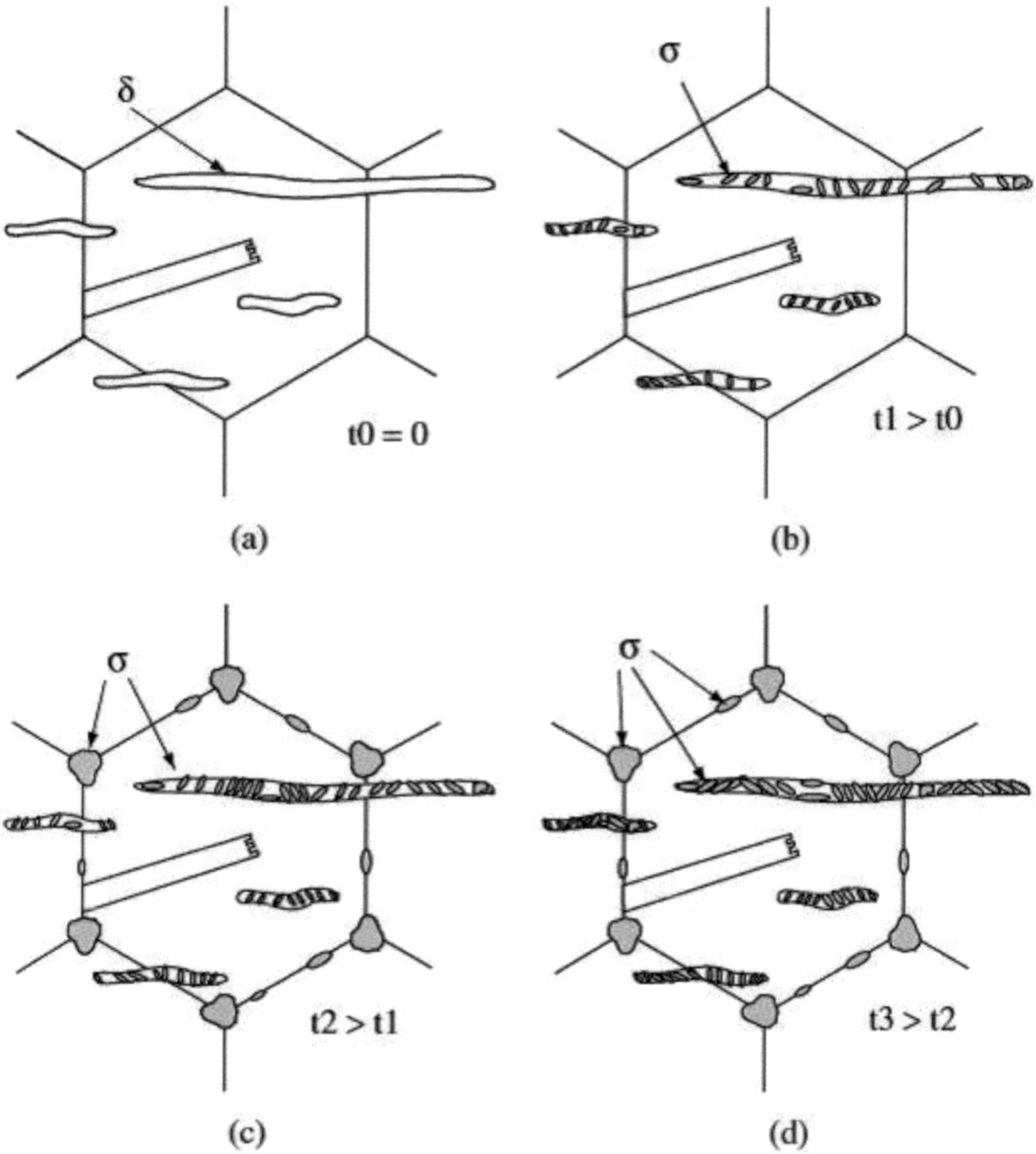

Figure 3. Precipitation mechanisms of σ-phase in AISI 316 L stainless steel [18].

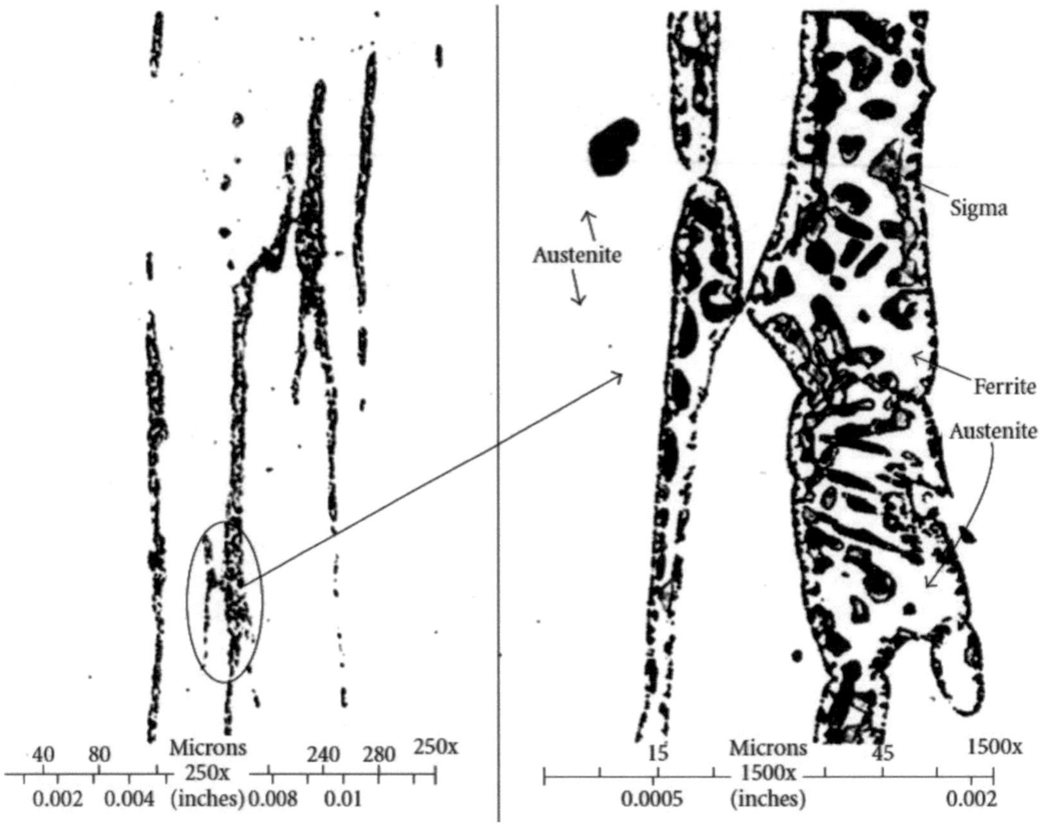

Figure 4. Microstructural observation of the $\delta \rightarrow \sigma + \gamma_2$ phase transformation in AISI 304 stainless steel (595°C, 31,000 h) [15].

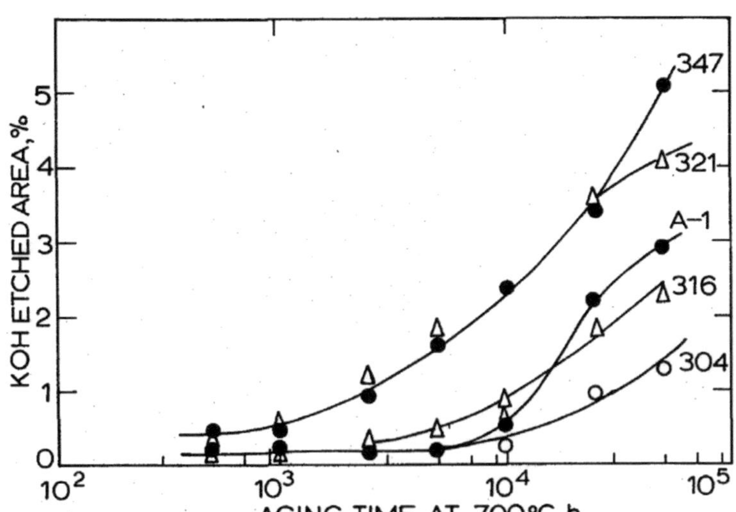

Figure 5. Precipitation of σ-phase in different grades of austenitic stainless steels [23].

According to Ref. [23], the behavior of Tempaloy-A1, which is 18/10 steel stabilized with Nb, but its Nb/C ratio is only 1.86 compared to 17.40 in the steel 347, supports the observation that σ-phase forms in steels with chromium equivalent higher than 18 wt.% when the carbon content falls below a critical value. In the 347, almost all the carbon is rapidly precipitated as NbC, while the low Nb content of Tempaloy-A1 means that a part of carbon is still dissolved in the solid solution. The different trend for 321 could be linked to the instability of TiC with regard to $M_{23}C_6$. The precipitation of $M_{23}C_6$ lowers both the carbon and the chromium contents.

The formation of σ-phase in austenitic stainless steel greatly depends on grain size, as it affects the density of nucleation sites [9]. According to Ref. [24] in austenitic stainless steels, both the grain size and grain shape influence sigma precipitation because grain boundaries represent high-energy nucleation sites for nucleation of σ-phase [25]. Fine-grained structure then provides more σ-phase formation.

The same is true for cold deformation, however, the effect of cold work on σ-phase precipitation depends mainly on its influence on recrystallization. The σ-phase formation is enhanced if the amount of deformation energy is sufficient to generate recrystallization at operating temperature. If the amount of cold work is too low and recrystallization does not occur, σ-phase formation can be actually retarded [26]. Hot working enhances σ-phase precipitation, as it was demonstrated on a duplex stainless steel hot deformed at 900°C. Hot deformation induces lattice defects which act as favorable sites for σ-phase precipitation. Higher total content of σ-phase in the specimen deformed at the lower strain rate ($0.01\ s^{-1}$) compared to the sample deformed at $1\ s^{-1}$ was explained by the difference in density and velocity of dislocations. Deformation at high strain rates results in higher dislocation density, which means precipitation of fine σ-phase particles throughout the whole specimen. These fine particles grow slowly, because they are in contact with individual dislocations for shorter time. As a result, material deformed at high strain rates has plenty of fine σ-phase particles but the total volume fraction of σ-phase is greater in samples deformed at lower strain rates [27].

4. Presence of sigma phase in the tube bends after high temperature exposure

Extensive experimental work confirmed the effect of cold bending on σ-phase precipitation. The study includes advanced steels for USC boilers. The effect of technology and high temperature exposure was evaluated (see **Figure 6**).

Series of tube bends were analyzed. The first part were cold worked bends without any further heat treatment (referred to as-bended state), the second part were heat-treated tubes after bending (referred to post-bend heat treatment—PBHT), third part of test program focused on evaluation of structural changes and changes in mechanical properties after exposure at high temperature.

The mechanical properties and fracture behavior were examined using miniaturized tensile tests and small punch test (SPT). The effect of plastic deformation applied during the bending of tubes on the microstructure and the mechanical properties were also studied. The positive effect of solution annealing on the microstructure was quantified as well [28, 29].

During the third part of the experimental program the tube bends exposed to high temperature in the boiler without any loading were investigated. The effect of high temperature exposure on formation of σ-phase and fracture energy after exposure was evaluated.

Bend tubes Ø 38 × 6.3 of Grades TP347HFG, Super 304H and HR3C and the bend radii R60, R80 and R100 were tested (see **Tables 1** and **2**). The samples of the tube bends were cut out from bended part of tubes (see **Figure 7**). The samples were then evaluated in as-received conditions (with PBHT and without PBHT) and installed into a boiler without any loading for 1 year at temperature ranges from 635 to 695°C and 726 to 775°C, respectively.

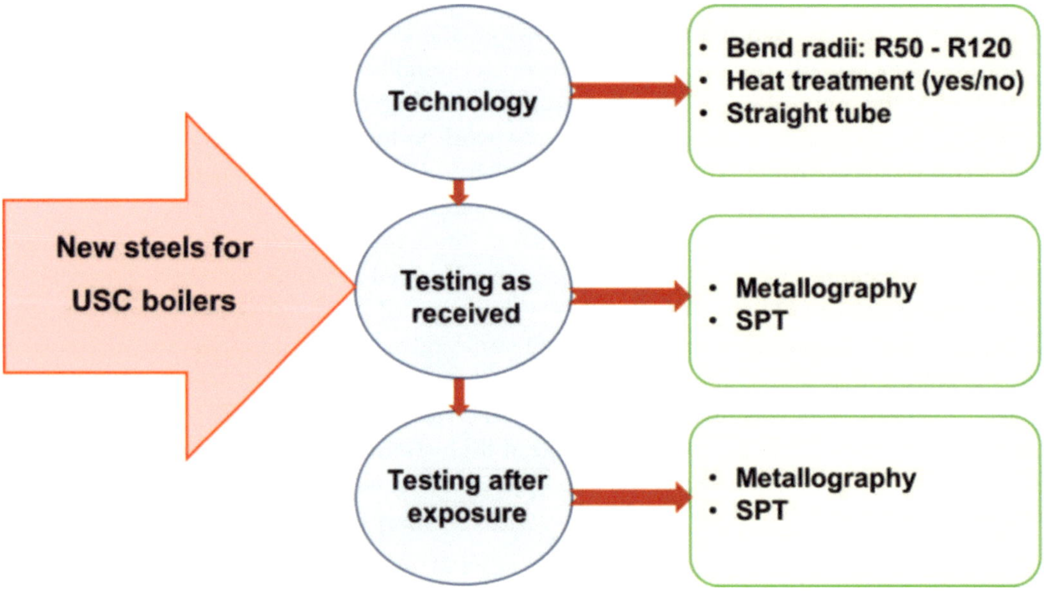

Figure 6. Diagram of evaluation of tube bends.

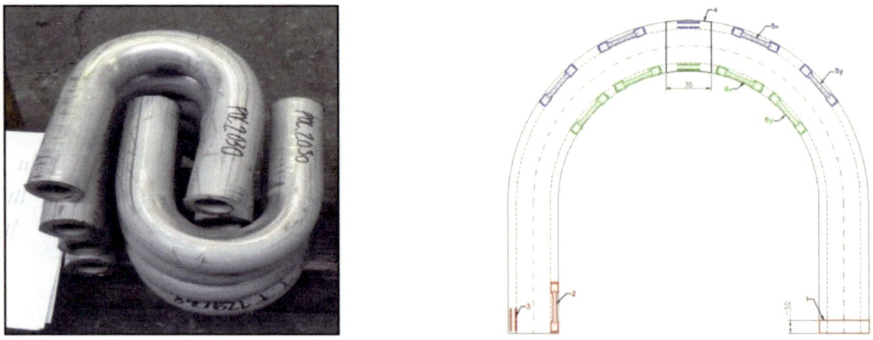

Figure 7. Samples in as-bended state under study and cutting plan.

5. Testing method used to evaluate mechanical properties of tube bends

In order to understand the effect of σ-phase on material properties and microstructure, various testing methods based on miniaturized test techniques have been used. As the standard tensile test is not sensitive enough to quantify the effect of σ-phase, a special testing method, small punch test (SPT), originally developed for residual lifetime assessment of power plants components, has been used. This testing method is covered at this time by draft of ISO standard. A basic tenet of small punch test method is to penetrate small volume of experimental material (**Figure 8a**). The small punch testing technique utilizes a small disc specimen, 8 mm in diameter and 0.5 mm in thickness, clamped around its circumference and indented by a spherical punch up to failure [30]. Monotonic load versus displacement records are used to correlate the tensile and fracture toughness parameters (see **Figure 8b**). Moreover, the fracture energy can be calculated as the area under the deflection curve. All tests presented in this work were performed at room temperature.

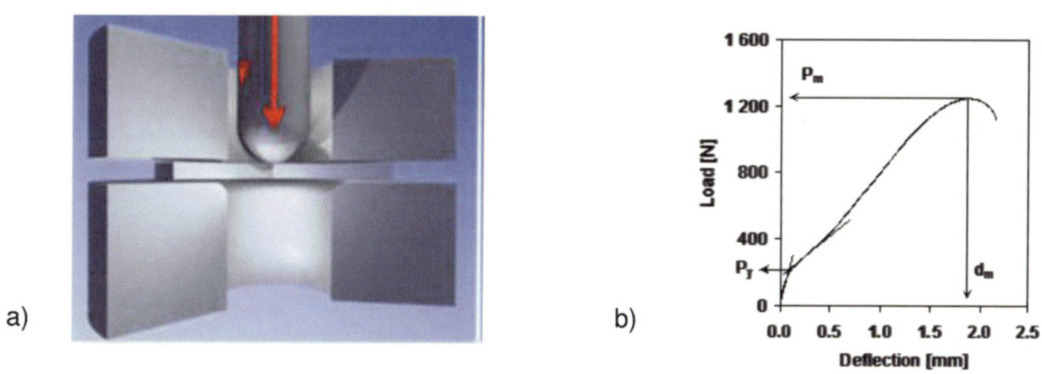

Figure 8. Principle of small punch test (a) and typical SPT record (b).

6. Results of metallographic analyses of tube bends after exposure at high temperature

Comparison of the structure of specimen in the as-bended state after exposure at lower and higher temperatures in the boiler for all three type steels can be seen in **Figures 9–11**.

Microstructural investigation showed that all tube bends that were exposed at the elevated temperature in the boiler without any loading, contain various amount of σ-phase. Amount of σ-phase depends on temperature of exposure, bend radii and heat treatment after bending. Therefore, the maximum amount was found in bends with smaller radii and without PBHT. The tube bends with smaller bending radii without PBHT, corresponds to maximum amount of cold work deformation, contain the largest amount of σ-phase for all grades. As amount of cold work deformation decrease, lower percentage of σ-phase can be observed, for example, in the structure of bend with higher radii where minimum amount of σ-phase was measured in the tube bends with bending radius R100 (**Figures 9–11**).

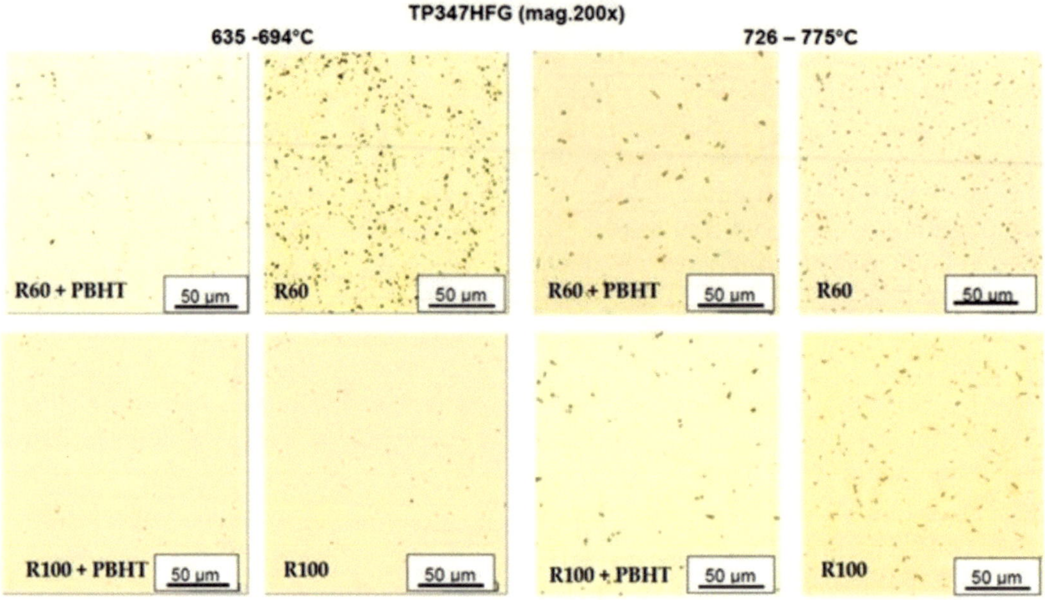

Figure 9. Comparison of the structure of specimen TP347HFG in the as-bended state after exposure at lower and higher temperatures in the boiler.

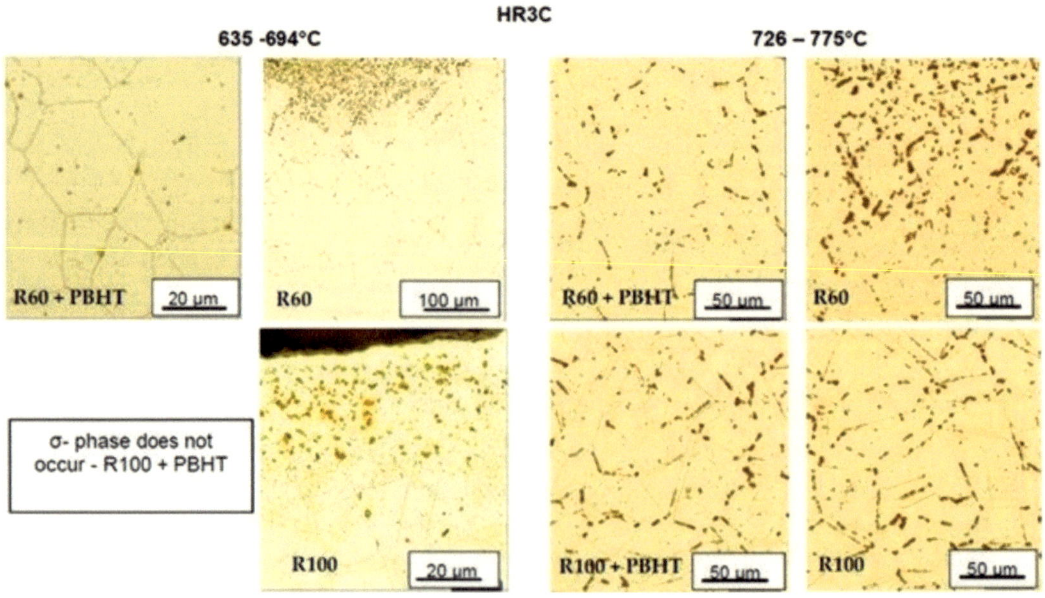

Figure 10. Comparison of the structure of specimen HR3C in the as-bended state after exposure at lower and higher temperatures in the boiler.

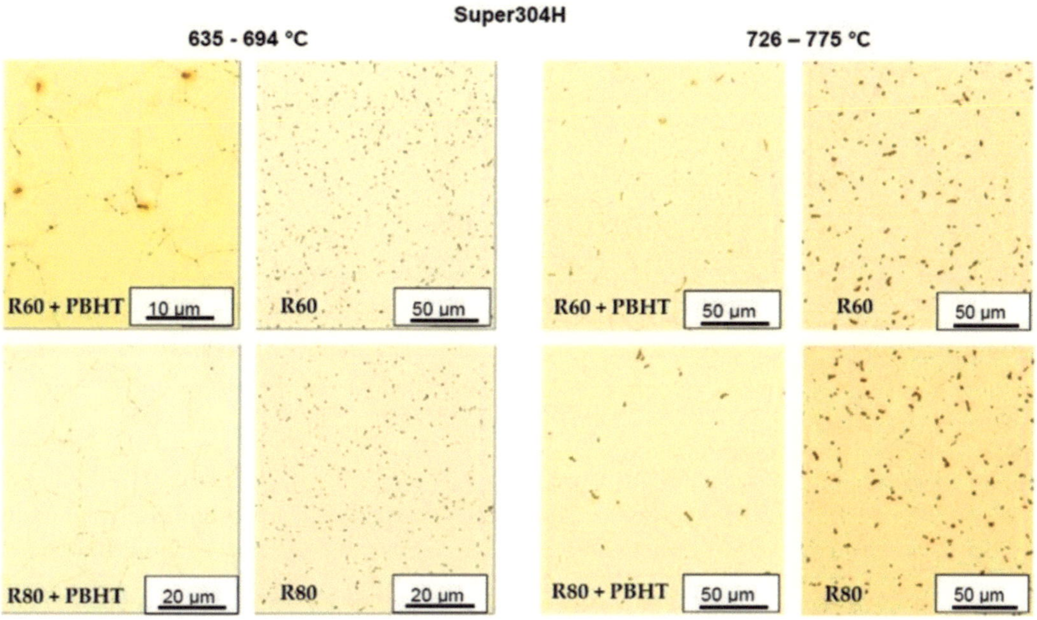

Figure 11. Comparison of the structure of specimen Super 304H in the as-bended state after exposure at lower and higher temperatures in the boiler.

Logically, tube bends exposed at higher temperature contain significantly higher amount of σ-phase, this fact is in very good agreement with theoretical results. The particles of σ-phase differ in quantity and in size and even in the distribution in the wall of the bent tube. Tube bends exposed at the lower temperature contain σ-phase mostly near the outer surface of extrados, corresponding to the position of the highest cold work deformation. While specimens exposed at higher temperature contain σ-phase in the whole wall of extrados.

Reciprocally, **Figure 10** shows the structure of HR3C tube bend after the exposure in two locations in the boiler. The tube bend with post-bend heat treatment (PBHT) exposed at lower temperature showed very few particles of σ-phase [31].

Tube bends without PBHT contain considerably more σ-phase. Higher temperature of exposure caused the coarsening of σ-phase in the tube bends that will significantly affect toughness of materials.

7. Effect of sigma phase on fracture energy

As demonstrated in **Figure 12**, the presence of even very small volume of σ-phase in the structure means drop in fracture energy and can be clearly detected from force-displacement record of the small punch test. The test results in as-bended state and after the exposure (at lower temperature) and the test results for samples with and without PBHT were compared. **Figure 12**

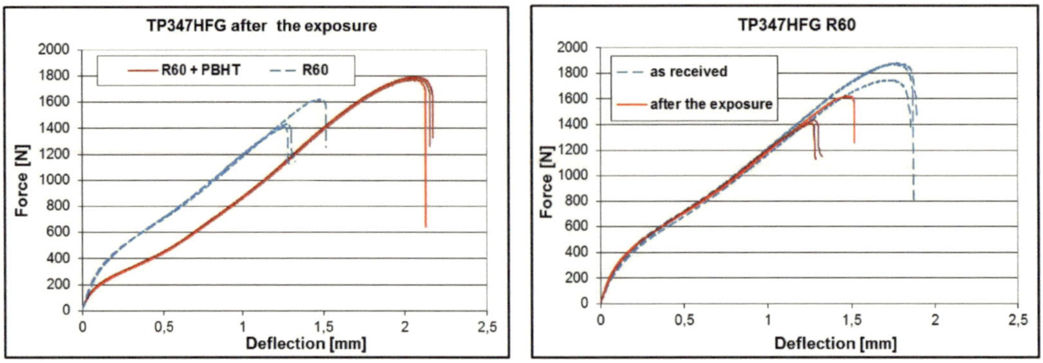

Figure 12. Force versus displacement of tube bend TP347HFG in as-bended state and after the exposure.

shows the effect of σ-phase on fracture energy represented by SPT method in TP347HFG samples and effect of PBHT for the same steel. It is clearly seen positive effect of PBHT on fracture energy as well as effect of exposure, where the fracture energy is decreased significantly.

Figure 13 shows effect of exposure on the fracture energy of HR3C steel for as-bended and PBHT state, respectively. In both examples was fracture energy significantly reduced by the high temperature exposure. In the case of as-bended tubes, changes in the slope of SPT records were identified as well. These changes can probably be connected with submicrostructural changes during exposure in this type of steel and will be studied in the future work.

Only moderate changes were identified in the case of Super 304H grade steel, see **Figure 14**. Drop in fracture energy is very small and no changes in the slope were recognized.

The obtained results indicate that σ-phase is one of important factors that should be remembered during high temperature exposure of these materials in USC blocks. The presence of σ-phase only at the outer surface of extrados at lower exposure temperatures and in the whole wall section of extrados at higher exposure temperature indicates, that the amount of σ-phase in the tube bends is also connected to the deformation energy introduced into the wall of

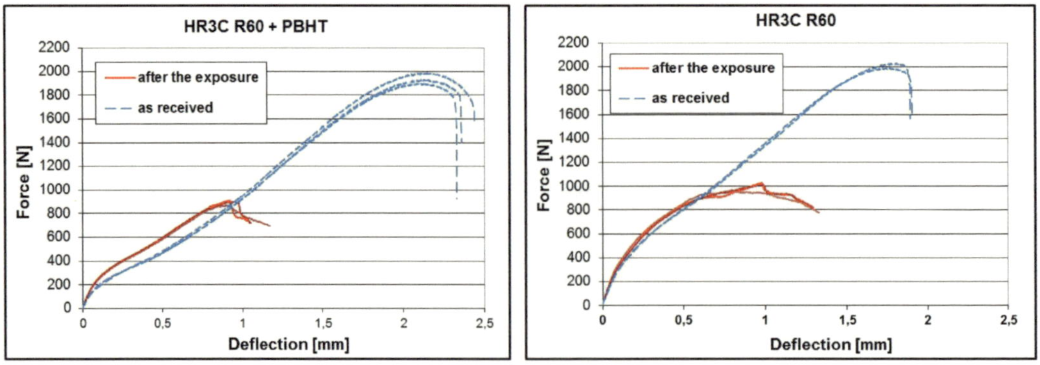

Figure 13. Force versus displacement of tube bend HR3C in as-bended state and after exposure.

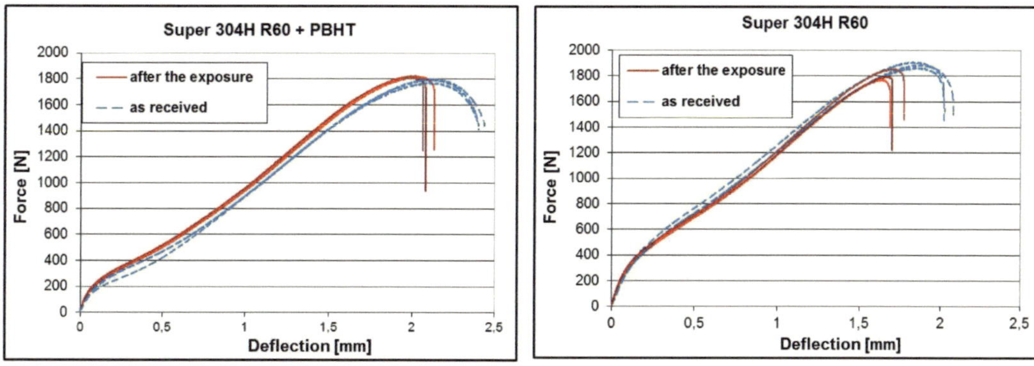

Figure 14. Force versus displacement of tube bend Super 304H in as-bended state and after exposure.

bend and the deformation energy seems to decrease the activation energy of σ-phase formation. Therefore, samples with PBHT and samples with large bend radii and relaxed or lower internal stresses, show only small, if any, content of σ-phase.

8. Importance of heat treatment of tube bends—when is HT necessary?

Based on experimental investigations mentioned above the tube bends with low bending radii (R60, R80) require heat treatment after bending to minimize internal residual stresses and tendency for precipitation of σ-phase in the structure. If the exposure temperature is higher than 700°C, it is necessary to apply heat treatment after bending even for bending radii R100.

After exposure in temperature range 635–694°C small amount of σ-phase was found in all specimens near the surfaces. After exposure in temperature range 726–775°C σ-phase was found across the whole wall thickness of extrados. In order to minimize the volume of σ-phase in the structure of extrados with small bend radii, it is necessary to carry out solution annealing after bending [28].

9. Degradation of material properties due to sigma phase in power plant tubes after 100,000 h of exposure

Very little data describing long-term exposure of TP347HFG and TP347H is available in the literature. Although there are many articles on short-term laboratory exposure and initial experience in power plants. The Danish coal-fired ultra-super critical (USC) plant, Nordjyllandsværket, after 100,000 h of operation has focused on the state of materials, such as P 91 or TP347HFG used in the boiler. Tube sections were removed from more than 20 different locations in the boiler during the 2012 summer shutdown to assess the material conditions in the boiler. The boiler was designed as an USC once-through Benson tower type with double reheat and was commissioned in 1998 with 290 bar and 580/580°C steam data [32].

Tube	Location in the boiler	Material	σ-phase (%)		Temperature of exposure (°C)	ASTM grain size G
			0°	180°		
10	Outlet SH1	TP347HFG	—	—	585	7–9
11	Outlet RH2.2	TP347HFG	—	—	600	6–8
13	SH1 middle	TP347H	1.1	0.2	585	7–8
14	Outlet RH1.2	TP347HFG	1.6	0.6	590	7–8
15	Outlet SH2	TP347HFG	1.2	—	600	6–8
16	Inlet SH2	TP347HFG	1.4	0.2	620	7–9
17	Inlet RH1.2	TP347HFG	5.5	0.6	585/570	7–9
18	Inlet SH1	TP347H	0.8	—	535	8
19	Inlet SH1	TP347H	2.2	—	550/535	8–9
20	Outlet Screen	TP347HFG	1.1	0.2	570/555	5–8
21	Outlet Screen	TP347HFG	—	—	570/555	5–8

Table 3. Area fraction of sigma phase in TP347H and TP347HFG tubes [32].

The tubes have been investigated in order to document the residual wall thickness, fire-side corrosion and streamside oxidation rates and morphology and microstructure evolution after 100,000 h of exposure. This assessment has revealed the presence of σ-phase in superheater (SH) and reheater (RH) tubes.

It is well-known that the orientation of tubes in the boiler has a great influence on the microstructure. Therefore, the fire-side facing the hot flue gas was marked as position 0° and the opposite side was marked 180° (see **Table 3**). **Table 3** also shows amount of σ-phase identified using SEM in various tubes.

10. Mechanical testing

Tensile strength seemed to be quite comparable in all tubes regardless of the position, that is, regardless of the content of σ-phase. On the other hand, yield strength at the position 0° was lower in tubes that contained higher amounts of σ-phase (tubes 17 and 19) and also in tube 13, which contained small amount of σ-phase, but was operated at the higher temperature (585°C), see **Table 4**.

Although strength seemed not to be affected by σ-phase appearance, plastic properties, such as elongation and reduction of area, showed a significant response. The presence of σ-phase resulted in a higher loss of elongation and reduction of area. The measured values of elongation and reduction of area can be directly related to toughness of steel as σ-phase appears to change the fracture properties. Nearly all investigated tubes showed decrease of elongation and reduction of area (8–25% depending on the content of σ-phase) compared to tubes without σ-phase.

Tube	Specimen ID	Tensile strength	Yield strength	Elongation	Reduction of area
		(MPa)		(%)	
17	Position 180°	675	357	48.0	64.0
		684	389	43.3	59.6
	Position 0°	676	277	38.7	45.9
		678	286	38.7	45.5
19	Position 180°	696	343	47.3	73.2
		697	347	50.0	73.3
	Position 0°	685	300	42.7	55.4
		684	301	41.0	55.4
13	Position 180°	715	374	51.3	71.6
		716	379	50.3	71.5
	Position 0°	700	331	46.7	60.9
		700	337	44.7	60.9

Table 4. Results of tensile test of tube with maximum content of σ-phase.

Although the presence of brittle phase in the structure cannot be unambiguously indicated by tensile testing, it could be clearly identified by drop in fracture energy during Charpy-V notch testing. The presence of σ-phase in the test sample resulted in drop of impact energy in all tested tubes, **Figure 15**. Even though the absolute values of impact energy between thin-walled and thick-walled tubes cannot be directly compared due to various specimen size, the difference between position 0 and 180° is clearly visible. The only exception is the tube 14, which contains the smallest amount of σ-phase. The decrease of the impact energy due to high temperature exposure and σ-phase precipitation is illustrated also in **Figure 15**, where the last green columns represent the impact energy in TP347HFG tubes in the as-received condition [32, 33].

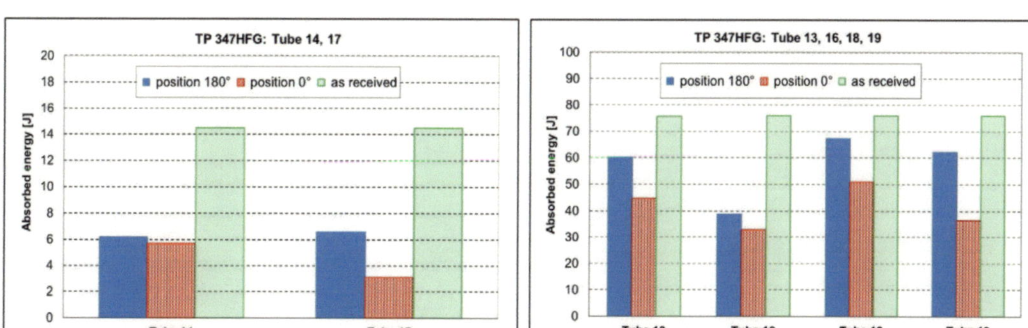

a) b)

Figure 15. Absorbed energy at the side with (position 0°) and without σ-phase (position 180°) and in as-received condition tubes with wall thickness 3.6 mm miniaturized Charpy V specimens 5 × 3 × 32 mm (a) and tubes with wall thickness higher than 7 mm – subsize Charpy-V specimens 10 × 5 × 55 mm (b).

11. Metallographic and fractographic investigations of sigma phase

In order to get a general overview of the microstructural changes and to quantify the amount of σ-phase, all samples were investigated by light optical microscopy (LOM) and by scanning electron microscopy (SEM). ASTM grain size was also measured, see **Table 3**.

Larger amount of σ-phase was found in position 0°, that is, on the fire-side. The σ-phase free band of about 300–400 μm at the surface on the flue gas side was observed. The greatest concentration of σ-phase was found in the mid-thickness of the tubes with decreasing concentration toward the inner surface of the tubes. Small amount of σ-phase was observed also in position 180° that is partially shielded from the hot flue gas by surrounding tubes [25]. **Figure 16** shows the difference in σ-phase content in position 0 and 180° for sample with the maximum content of σ-phase.

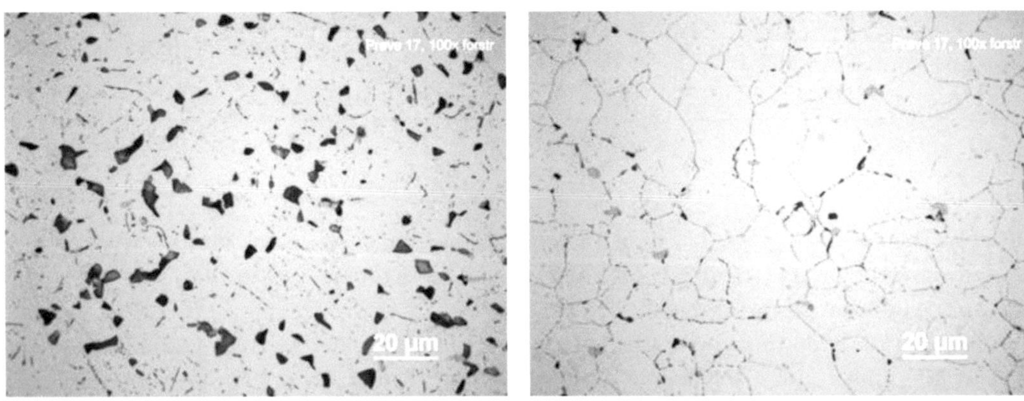

a) b)

Figure 16. Tube17: position 0°—5.5% of sigma phase (a), position 180°—0.6% of σ-phase (b).

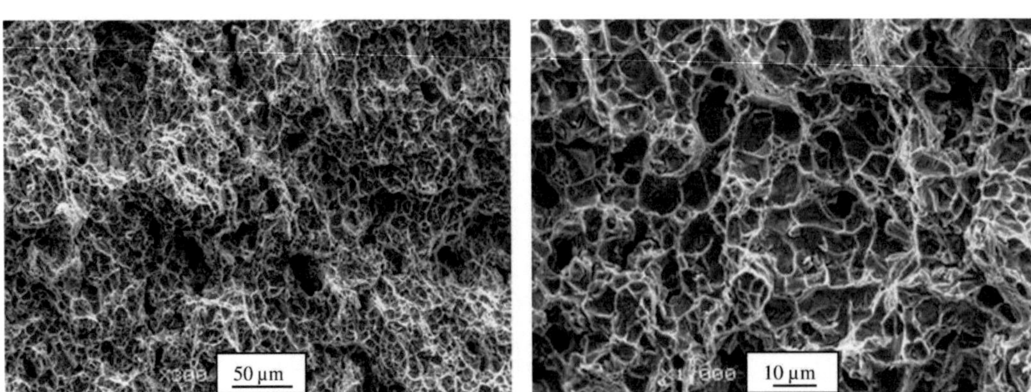

a (300x) b (1000x)

Figure 17. Tube 17: position 0° (high content of σ-phase) combined ductile/brittle fracture mode (a) and detailed image of brittle cleavage facets connected by ductile tearing ridges (b).

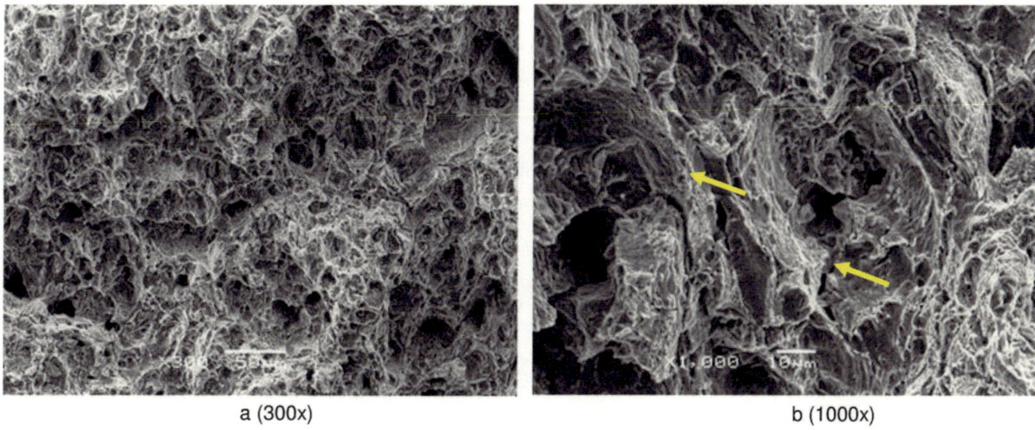

Figure 18. Tube 17: position 180° (without σ-phase) combined ductile/brittle fracture mode (a) and detailed image of the transgranular ductile fracture and holes with particles (marked with red arrows) (b).

Drop of fracture energy due to σ-phase has become evident on the fracture surface. The fractographic analysis performed on the surfaces of miniaturized Charpy specimens revealed significant differences in the morphology of fracture surfaces and these differences qualitatively corresponded to different values of impact energy observed in both cases (**Figures 17** and **18**).

12. Small punch testing

Another way how to cope with the fact that tensile test cannot respond to σ-phase presence in the microstructure is application of small punch test method. **Figure 19** shows comparison of test records (force versus displacement) of the SPT specimens from both sides of the tube 17 containing the highest amount of σ-phase. Five SPT specimens were prepared from each side of tube in order to obtain the information about the scatter of results. The difference between the side with and without σ-phase can be imagined as the area under the curves representing fracture energy. The drop of fracture energy between both sides of one tube was about 50%.

The values of fracture energy of all tested tubes are shown in **Figure 20**, where the test results are complemented by the values of fracture energy of TP347HFG steel in as-received condition [32]. In this way, it was possible to compare both the effect of σ-phase and the effect of operation conditions on fracture behavior of the steel under investigation. It was assumed that the difference between position 0 and 180° represented the effect of σ-phase in the structure after long-term operation at high temperature.

The surprising result of this testing was the fact that the fine-grained tubes (14, 16, 17) are more sensitive to the presence of σ-phase in the structure than coarse-grained tubes (13, 18, 19). It seems that the coarse-grained tubes are more resistant to embrittlement effect of σ-phase than the fine-grained tubes. This result is in great contradiction to the widely published statement about the higher embrittlement of coarse-grained material due to σ-phase precipitation.

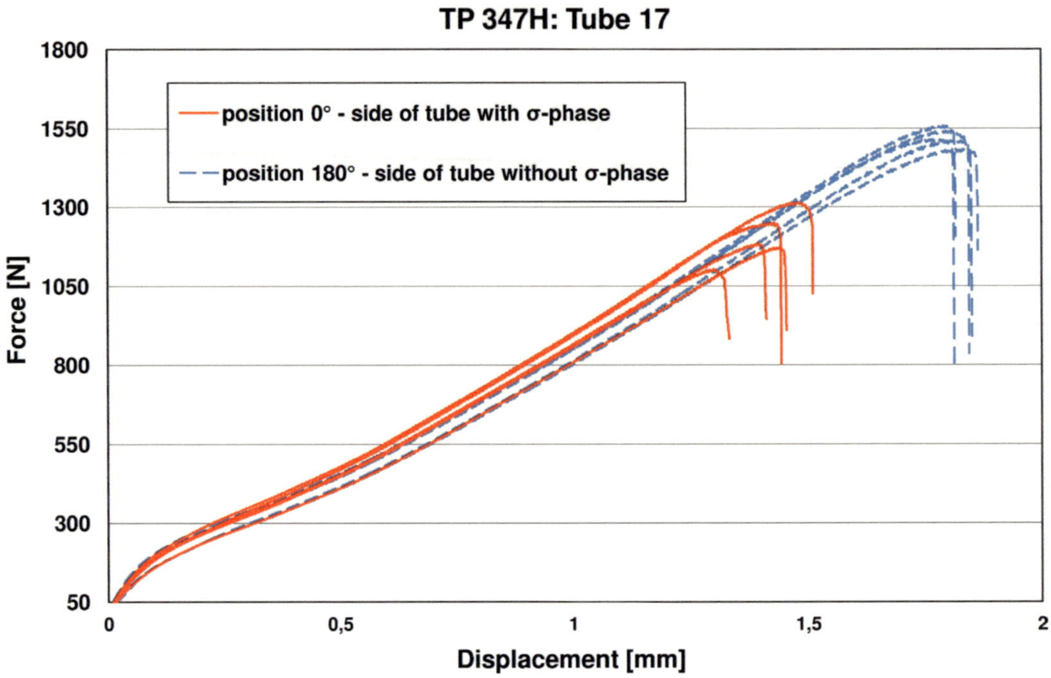

Figure 19. Comparison of absorbed fracture energy during SPT testing of tube 17 at position 0 and 180°.

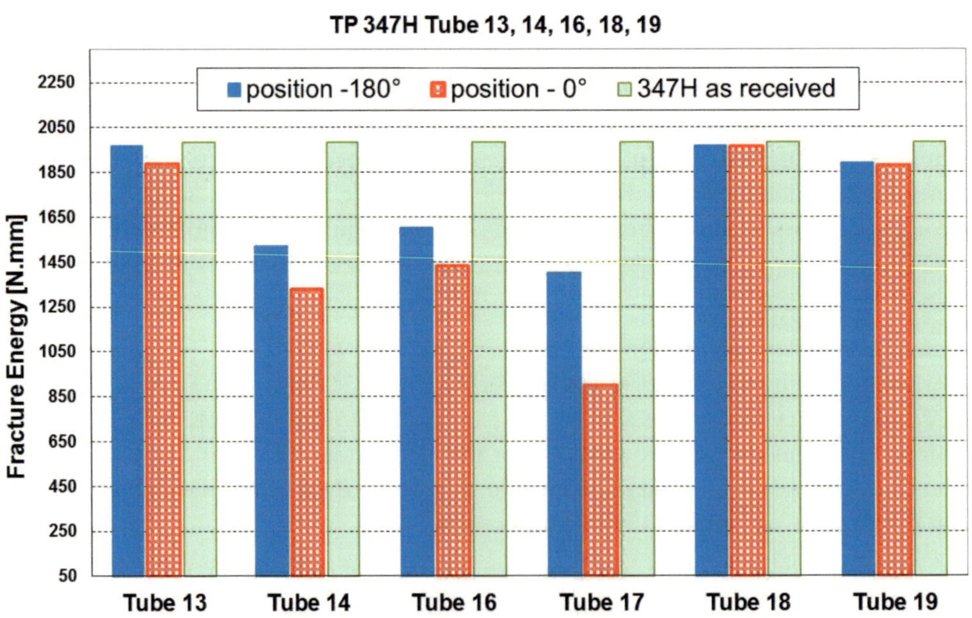

Figure 20. Fracture energy of tube position 0 and 180° and as-received state.

http://dx.doi.org/10.5772/intechopen.71569

All the performed experiments thus confirmed that σ-phase precipitated in the tube bends of advanced heat-resistant austenitic steels for USC conditions after 1 year of high temperature exposure without any external loading. The strong embrittlement effect of σ-phase was clearly confirmed by drop of fracture energy of small punch tests. Cold work, which stores the deformation energy in the microstructure, accelerates the precipitation of σ-phase. However, this effect can be eliminated by the post-bend heat treatment.

The negative effect of sigma phase on fracture energy of TP347HFG can clearly be seen from results obtained by evaluation of microstructure and mechanical properties of superheater tubes made of TP347HFG grade after exposure high temperatures for 100,000 h in a USC plan, where the amount of σ-phase was measured to be about 5%.

13. Creep properties and structural changes of base material and welded joints of advanced austenitic steels

Tubes Ø 38 × 6.3 mm made from steels Super 304H and HR3C were used for the evaluation of creep properties of the tubes and their welded joints. The circumferential tube joints were performed by GTAW (method 141) automated orbital welding (position PK), where both ends of the tube are centered in horizontal position and do not move during welding. Generally, welding of all austenitic steels must be performed with restricted thermal input in order to avoid hot cracking of weldments. Heat input for individual layers was therefore in the range from 1100 to 1900 J/mm and interpass temperature was kept at 150°C. Welding wires Ø 0.8 mm were used for welding. A matching filler metal Thermanit 304HCu was used for homogeneous weld joints of Super 304H steel, whereas overmatching nickel-based alloy filler metals Thermanit 617 and UTP A6170Co were used for homogeneous weld joint of HR3C steel. The chemical compositions of the filler metals are shown in **Table 5** [34].

Stress rupture tests of both base materials and welded joints performed at temperatures 650, 700 and 750°C. The calculated values of 10,000 h ($R_{u/T/10^4}$) creep rupture strength are stated in **Table 6**. The calculation was performed using Seifert parametric equation [35]:

$$\log R_{uT} = A_0 + A_1 P + A_2 P^2, \quad where \ P = \left[T(C + \log t_r)\right] 10^{-4} \qquad (2)$$

where T is the absolute temperature in K, t_r is time to fracture in hours, A_1, A_2, A_k and C are material constants.

Filler metal	C	Si	Mn	Cu	Mo	Cr	Ni	Ti	Fe	Nb	Al	N	Co
Thermanit 304HCu	0.10	0.4	3.2	3.0	0.8	18.0	16.0	—	—	0.4	—	0.2	—
Thermanit 617	0.05	0.1	0.1	—	9.0	21.5	rest	0.5	1.0	—	1.0	—	11.0
UTP A 6170 Co	0.06	0.3	—	—	8.5	22.0	rest	0.4	1.0	—	1.0	—	11.5

Table 5. Nominal chemical composition of filler materials, wt.% [34].

Steel	Super 304H			HR3C		
Temperature	650°C	700°C	750°C	650°C	700°C	750°C
Tube	174	104	60	168	104	62
Weldment	155	95	60	141	99	70
Datasheet	160	101	61	174	106	64
SRF (−)	0.89	0.91	1.00	0.84	0.95	1.00

Table 6. Creep rupture stress $R_{u/T/10^4}$ of the Super 304H and HR3C steel tubes and weldments, MPa.

Table 6 shows also the results of calculation of strength reduction factor (SRF) expressing the ratio of creep rupture strength of welded joints $R_u(W)$ and the creep strength of base metal $R_u(BM)$, that depends on the temperature and time to rupture [36]:

$$SRF = \frac{R_u(W)}{R_u(BM)} = f(t, T) \leq 1 \tag{3}$$

Recalculation of experimental results into the value of Larson-Miller parameter P_{LM} was used to compare the evaluated creep resistance of the base metal and welded joint in the Eq. (4) [37]:

$$P_{LM} = T[\log(t) + C] \tag{4}$$

with the usual meanings of the variables. Constants C equals to 20.8 in Super 304H and 16.8 in HR3C, which are the values calculated using least squares method from stress rupture data stated in material sheets [38, 39].

Creep rupture strength as well as creep behavior of both steels is very similar, their strength reduction factor increases with increasing temperature and at 700°C creep rupture strength of welded joint overcomes that for base metal and SRF equals to one. Similar behavior was observed also in other tested welded joints of Super 304H steel [40]. Literature data show that and this type of behavior is typical for nearly all austenitic heat-resistant steels [41–43].

The comparison of creep tests of the tubes and weldment of Super 304H and HR3C steels is shown in **Figure 21**, base metal (dark symbols) and welded joints (light symbols). Still running stress rupture tests are marked by red arrows. The solid lines labeled WB 550 and WB 47 represent the average creep rupture strength of steels Super 304H according to Ref. [38] and HR3C according to Ref. [39], the dashed lines then correspond to the lower −20% tolerance limit. All the results confirmed good creep resistance of the tested tubes as well as their weld joints.

However, the danger of σ-phase precipitation in the steel does not lie in lowering the creep strength but in exhausting of plasticity and embrittlement of the structure. This phenomenon can be clearly seen in **Figure 22**, showing the dependence of reduction of area on Larson-Miller parameter in both basic material and weldments.

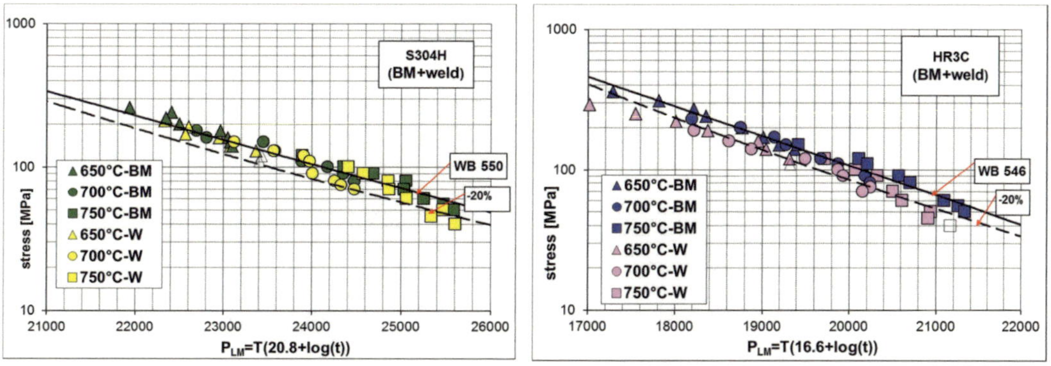

Figure 21. Dependence of stress on L-M parameter of Super 304H and HR3C steels and their weldments.

Plasticity drop in stainless steels at room temperature associated with precipitation of σ-phase is a well-known phenomenon. However, σ-phase seems to precipitate in shorter time and the drop of reduction of area seems to be deeper than that observed in austenitic heat-resistant steels of 304, 316, 321 and 347 types formerly used for the outer parts of superheaters of supercritical boilers.

The structural change and the form and extent of the σ-phase precipitation during creep exposure can be demonstrated on the example of above mentioned circumferential welds on tubes made of HR3C steel where structural states are compared here, in the as-welded state and after creep exposure at 750°C/50 MPa for 7704 h.

In the as-welded state the microstructure of the base metal of the welded joint close to the fusion line consisted of austenitic grains of various size (**Figure 23**). The microstructure of the nickel-based weld metal is documented in **Figure 24**. The austenitic structure contained relatively large primary carbides inside the grains, but no carbides were present at the grain boundaries of base material (**Figures 25** and **26**).

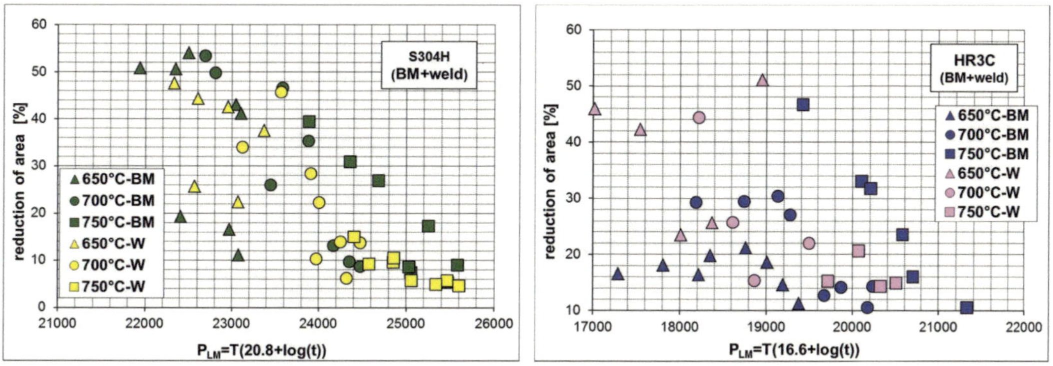

Figure 22. Dependence of reduction of area on L-M parameter of Super 304H and HR3C steels and their weldments.

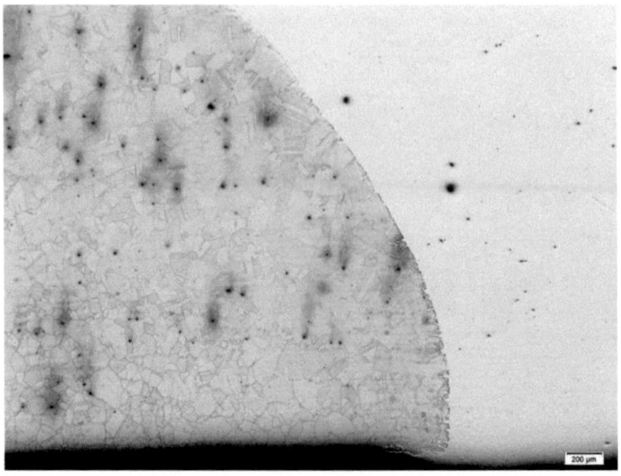

Figure 23. Microstructure of the root part of welded tube made of steel HR3C.

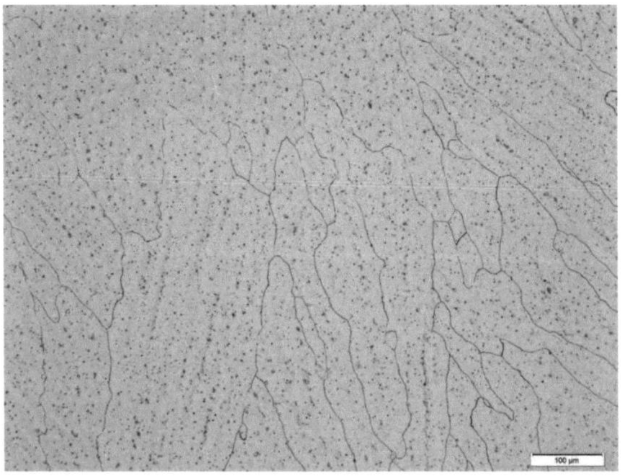

Figure 24. Microstructure of weld metal (Thermanit 617).

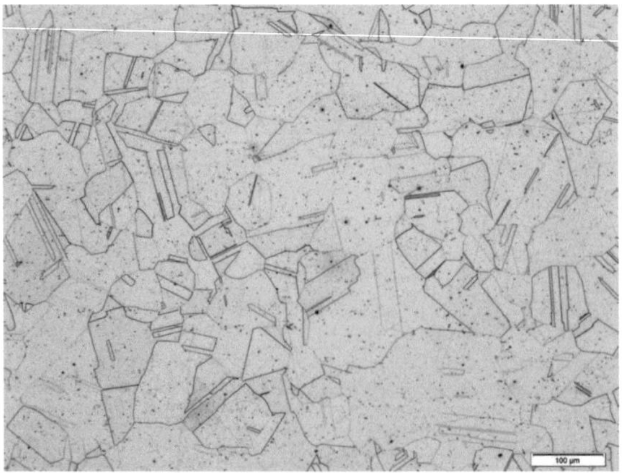

Figure 25. Microstructure of base material.

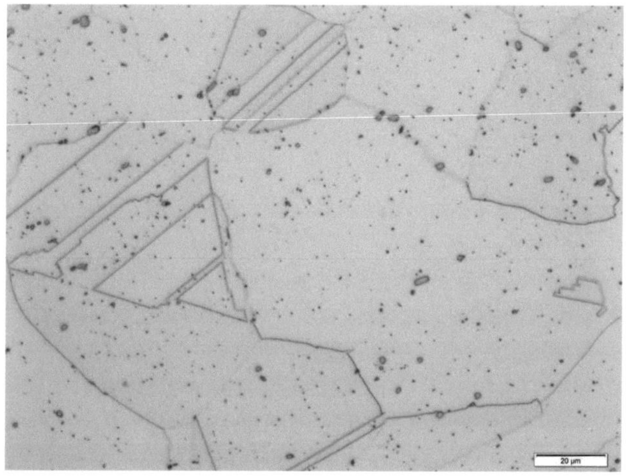

Figure 26. Detail of microstructure of base material.

The final rupture of the analyzed creep specimen started in the fusion line and propagated through the coarse-grained part of heat-affected zone (**Figures 27** and **28**). The failure was brittle, mostly intergranular and the reduction of area was very low, only 2.8%. Even though σ-phase precipitated in the base material, too, it formed randomly distributed particles that are not so dangerous as a network of σ-phase particles along grain boundaries detected in the coarse-grained HAZ, compare **Figures 29** and **30**.

The microstructural analysis thus confirmed the correlation between low plasticity of steel after long-term creep exposure at high temperatures and precipitation of σ-phase on grain boundaries of austenite. Especially in welded joints precipitation in the form of network on grain boundaries can lead to the type III cracking, that is, failure in the coarse-grained part of heat-affected zone close to the fusion line. This part of heat-affected zone is overheated during welding and accompanying grain growth will reduce the number of grains and the length of grain boundaries. Precipitating σ-phase along grain boundaries can form more or less continuous network and further reduce plasticity of this part of heat-affected zone that has due to its microstructure high, creep strength but very low toughness and plasticity.

Figure 27. Macrostructure of failure location of creep specimen tested at 750°C/7704 h.

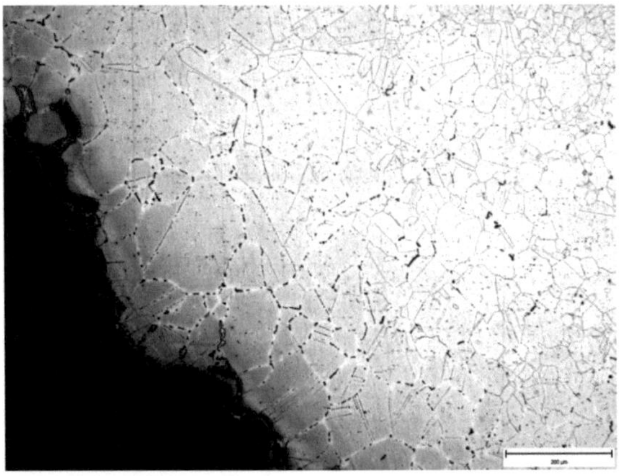

Figure 28. Microstructure of coarse-grained HAZ (etched in KOH).

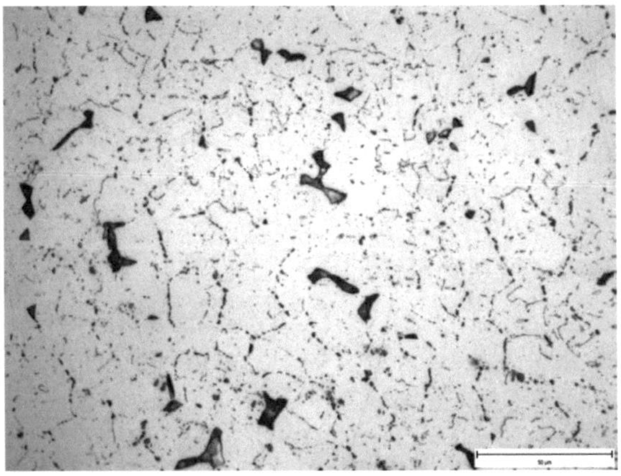

Figure 29. Detail of σ-phase particles in the base metal (etched in KOH).

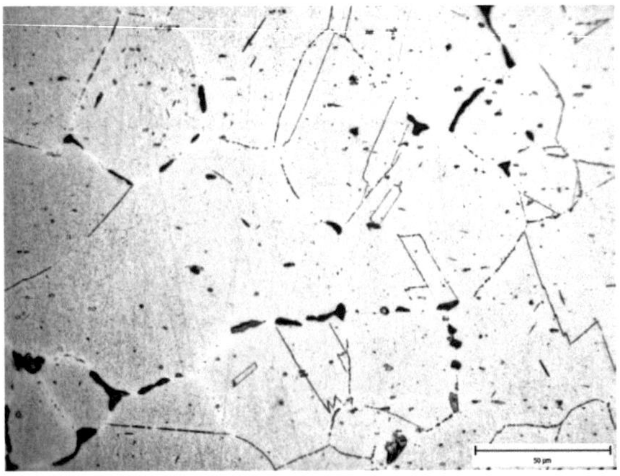

Figure 30. Detail of network of σ-phase along grain boundaries in coarse-grained HAZ (etched in KOH).

http://dx.doi.org/10.5772/intechopen.71569

14. Conclusions

High temperature and long-term exposure of all heat-resistant steels in USC and advanced ultra-super critical (A-USC) boilers brings about advanced demands for their structural stability. Increased operating temperatures accelerate all degradation processes that appear in all types of heat-resistant steels, from low-alloy through chromium modified toward austenitic stainless steels. Precipitation of σ-phase represents one of the most potentially dangerous degradation mechanism in advanced austenitic heat-resistant steels. Although precipitation of this phase was observed after long-term exposure at high temperatures in many austenitic steels, the results obtained during testing of superheater tubes and their bends as well as welded joint have shown that the precipitation of σ-phase starts in these steel grades after few thousands of high temperature exposure and even without external loading. Stress and plastic deformation induced in the material during cold bending and/or welding can further accelerate the commencement of its precipitation. Although there are developed also steels intentionally precipitation strengthened by σ-phase [44], the results obtained on three advanced austenitic heat-resistant steel grades Super 304H, HR3C and TP347HFG have proved the detrimental effect of σ-phase precipitation namely on toughness and plasticity of these steels.

Acknowledgements

This chapter was created in the Project No. LO1203 "Regional Materials Science and Technology Centre—Feasibility Program" funded by the Ministry of Education, Youth and Sports of the Czech Republic.

Author details

Zdeněk Kuboň*, Šárka Stejskalová and Ladislav Kander

*Address all correspondence to: creep.lab@mmvyzkum.cz

Material & Metallurgical Research, Ltd., Vítkovice, Ostrava, Czech Republic

References

[1] Viswanathan R, Purgert R, Rao U. Materials Technology for Advanced Coal Power Plants. Proceedings of the 1st Int. Conference on Super-High Strength Steels. Associazione Italiana di Metallurgia; Rome; 2005

[2] Masuyama F. Steam plant material developments in Japan. In: Lecomte-Beckers J et al., editors. Materials for Advanced Power Engineering, Part III. Forschungszentrum Juelich, Juelich; 1998. p. 1807-1824

[3] Larsen OH. Austenitic steels–mechanical properties, exfoliation, availability. In: Material and Quality Assurance. VGB Workshop; May 13-15; Copenhagen; 2009. p. Doc. no 541576

[4] Chi C, Yu H, Xie X. Advanced austenitic heat-resistant steels for ultra-super-critical (USC) fossil power plants. In: Alloy Steel-Properties and Use. Rijeka, Croatia: InTech; 2011. ISBN 978-953-307-484-9

[5] Duhaj P, Ivan J, Makovicky E. Sigma phase precipitation in austenitic steels. JISI. 1968; **206**:1245-1252

[6] Wilms ME, Gadgil VJ, Krougman JM, Ijsseling FP. The effect of σ-phase precipitation at 800°C on the corrosion resistance in sea-water of a high alloyed duplex stainless steel. Corrosion Science. 1994;**36**(5):871-881

[7] Souza CM, HFG A, SSM T, JMA R. The σ phase formation in annealed UNS S31803 duplex stainless steel: Texture aspects. Materials Characterization. 2008;**59**:1301-1306

[8] Sasikala G, Ray SK, Mannan SL. Kinetics of transformation of delta ferrite during creep in a type 316 (N) stainless steel weld metal. Materials Science and Engineering A. 2003;**359**(1):86-90

[9] Barcik J. Mechanism of σ-phase precipitation in Cr–Ni austenitic steels. Materials Science and Technology. 1988;**4**:5-15

[10] Hsieh CC, Wu W. ISRN overview of intermetallic sigma phase precipitation in stainless steels. Metallurgy. 2012;**2012**

[11] Woodyatt LR, Sims CT, Heattie HJ. Prediction of sigma type phase occurrence from compositions in austenitic superalloys. Transactions of the Metallurgical Society of the American Institute of Mechanical Engineers. 1966;**236**:519-527

[12] Guan K, Xu X, Xu H, Wang Z. Effect of aging at 700°C on precipitation and toughness of AISI 321 and AISI 347 austenitic stainless steel welds. Nuclear Engineering and Design. 2005;**235**:2485-2494

[13] Liu F, Hwang YH, Nam SW. The effect of post weld heat treatment on the creep-fatigue behavior of gas tungsten arc welded 308L stainless steel. Materials Science and Engineering A. 2006;**427**:35-41

[14] Chen TH, Yang JR. Effects of solution treatment and continuous cooling on σ-phase precipitation in a 2205 duplex stainless steel. Materials Science and Engineering A. 2001;**311**:28-41

[15] Restrepo Garcés G, Le Coze J, Garin JL, Mannheim RL. σ-Phase precipitation in two heat-resistant steels—Influence of carbides and microstructure. Scripta Materialia. 2004; **50**(5):651-654

[16] Redjaimia A, Metauer G, Gantois M. Decomposition of Delta ferrite in a Fe-22Cr-5Ni-3Mo-0.03C duplex stainless steel. A morphological and structural study. In: Proceedings of Duplex Stainless Steels, Vol. 1; Beaune, France. 1991. p. 112-126

[17] Wasnik DN, Dey GK, Kain V, Samajdar I. Precipitation stages in a 316L austenitic stainless steel. Scripta Materialia. 2003;**49**:135-141

[18] Villanueva DME, Junior FCP, Plaut RL, Padilha AF. Comparative study on sigma phase precipitation of three types of stainless steels: Austenitic, superferritic and duplex. Materials Science and Technology. 2006;**22**(9):1098-1104

[19] Shinohara K, Seo T, Kumada K. Recrystallization and sigma phase formation as concurrent and interacting phenomena in 25%Cr-20%Ni steel. Materials Transactions. 1979; **20**(12):713-723

[20] Baerlecken E, Fabritius H. Umwandlungskinetik der sigmaphase in einer Eisen-Chrom-Legierung mit 48% Chrom. Arch Eisenhuttenwes. 1955;**26**:679-686

[21] Garzón CM, Ramirez AJ. Growth kinetics of secondary austenite in the welding microstructure of a UNS S32304 duplex stainless steel. Acta Materialia. 2006;**54**(12):3321-3331

[22] Weiss B, Stickler R. Phase instabilities during high temperature exposure of 316 austenitic stainless steel. Acta Metallurgica. 1972;**3**(4):851-866

[23] Minami Y, Kimura H, Ihara Y. Microstructural changes in austenitic stainless steels during long-term aging. Materials Science and Technology. 1986;**2**:795-806

[24] Schwind M, Kallqvist J, Nilsson J-O, Agren J, Andren H-O. σ-phase precipitation in stabilized austenitic stainless steels. Acta Materialia. 2000;**48**:2473-2481

[25] Sourmail T, Bhadeshia HKDH. Modelling simultaneous precipitation reactions in austenitic stainless steels. CALPHAD. 2003;**27**:169

[26] Joseph R. Davis Ed., editor. ASM Specialty Handbook: Heat-Resistant Materials. ASM International, Metals Park, Ohio. ISBN 0871705966; 1997. 591 p

[27] Chandra T, Kuchlmayr R. Effect of strain rate on sigma formation in ferrite-austenite stainless steel at high temperatures. Journal of Materials Science. 1988;**23**:723-728

[28] Stejskalová Š, Kander L, et al. Study of Mechanical Properties and Microstructure Of Austenitic Stainless Bend Tubes, T-44/2014 Technical Report, Ostrava September; 2014 (in Czech)

[29] Kander L, Stejskalová Š, Čížek P. The change of the structure and mechanical properties of the austenitic steels after exposure at the critical temperature. In: Conference Proceedings METAL 2016: 25th Anniversary International Conference on Metallurgy and Materials; 25.-27.5; Brno, Czech Republic, Ostrava, Czech Republic, EU: TANGER Ltd., ISBN 978-80-87294-67-3; 2016. p. 724-729

[30] Small Punch Test Method for Metallic Materials. Cen Workshop Agreement CWA 15627, 2007

[31] Stejskalová Š, Kander L, Hermanová Š. The change of the structure and mechanical properties of the austenitic steels after exposure at the critical temperature. In: Longauerová M, Horňák P, Zubko P, editors. Metallography XVI: 16th International Symposium on Metallography and Materials Science; April 20-22; Stará Lesná, Slovakia. Switzerland: Trans Tech Publications; 2016. p. 330-334

[32] Korčáková L, Kander L, Montgomery M, Jensen HT, Stejskalová Š. The influence of sigma phase precipitation on the mechanical properties of TP 347H austenitic steels after 100,000. In: Matocha K, Hurst R, Sun W, editors. The 3rd International Conference SSTT 2014: Determination of Mechanical Properties of Materials by Small Punch and Other Miniature Testing Techniques; 2014. Castle Seggau, Austria: OCELOT, ISBN 978-80-260-6722-1; 2014. p. 119-132

[33] Kander L. Evaluation of Effect of Bending on Structure and Mechanical Properties of Austenitic Steels Super 304H, HR3C and TP347HFG using Small Punch Tests, T-63/2011, Materiálový a metalurgický výzkum. Ostrava, T-11/2011 (in Czech)

[34] Pomikálek L, Hermanová Š, Dobrovodská L. Effect of welding on the properties of HR3C, super 304H, Tp347HFG and P92 steels. In: Metal 2013: 22nd International Conference on Metallurgy and Materials; may 15-17; Brno, Czech Republic. Ostrava: Tanger; 2013. p. 806-811

[35] Seifert W. Statistische Kenngrössen aus der Ausvertung von Zeitstandversuchen. In: Warmfeste metallische Werkstoffe; Zittau. Kammer der Technik. 1987. p. 129

[36] Kimmins ST, Coleman MC, Smith DJ. An overview of creep failure associated with heat affected zone of ferritic weldments. In: Creep and Fracture of Engineering. Materials and Structures. London: The Institute of Metals; 1993. p. 681-694

[37] Larson FR, Miller J. Time-temperature relationship for rupture and creep stresses. Transactions of the ASME. 1952;**74**:735-775

[38] VdTÜV-Werkstoffblatt 550 - Warmfester Walz- und Schmiedestahl X10CrNiCuNb18-9-3, Düsseldorf : VdTÜV, 12/2010

[39] VdTÜV-Werkstoffblatt 546 - Warmfester Walz- und Schmiedestahl X6CrNiNbN25-20 (1.4952), Düsseldorf : VdTÜV, 12/2009

[40] Tortorelli PF, Unocic KA, Wang H, Santella ML, Shingledecker JP. Alloys for advanced Ultrasupercritical (A-USC) steam boilers. In: EPRI. Crosscutting Research Materials Program DOE Fossil Energy; April 20; Pittsburgh, Pennsylvania. 2016

[41] Price AT, Williams JA. Influence of welding on creep properties of steels. In: Wilshire B, editor. Recent Advances in Creep and Fracture of Engineering Materials and Structures. Swansea: Pineridge Press; 1982. p. 265-353

[42] Etienne CF, Heerings JH. Evaluation of the influence of welding on creep resistance. Steel Research. 1995;**65**:187-196

[43] Sobotka J, Bobek J, Sobotková M. Long-term creep strength and SRF for welded joints of 316H steel boiler tubes. In: Stainless Steel 96. Düsseldorf: VDEh; 1996. p. 363-367

[44] Di Gianfrancesco A, editor. Materials for Ultra-Supercritical and Advanced Ultra-Supercritical Power Plants. Duxford: Woodhead Publishing; ISBN 9780081005583; 2016. 900 p

Chapter 5

Precipitation Processes in Creep-Resistant Austenitic Steels

Grzegorz Golański, Adam Zieliński and
Hanna Purzyńska

Additional information is available at the end of the chapter

http://dx.doi.org/10.5772/intechopen.70941

Abstract

Creep-resisting austenitic steels constitute a group of construction materials which can work in the conditions of creep for the temperature range from 550 to 700°C. The service of austenitic steels leads to the progressive degradation of their microstructure, which results in the changes of functional properties. The main mechanisms of degradation of the austenitic steel microstructure include the processes of matrix softening, the processes of precipitation and matrix depletion of the interstitial and substitution elements. Precipitation processes in austenitic steels are a very important indicator, which allows the advancement of microstructure degradation processes in these steels to be determined. Hence, the knowledge of the impact of individual secondary phases on the microstructure and properties of austenitic steels plays a very important role in diagnosing the components and equipment of the power boiler system and makes it possible to forecast the time of safe operation of systems made from these steels. Based on own studies and data from literature, this paper will present the characteristics of secondary phase precipitates occurring in creep-resistant austenitic steels during their operation at an elevated/high temperature. The effect of secondary precipitates on mechanical properties of these steels will be discussed too.

Keywords: creep-resistant austenitic steel, precipitation processes, mechanical properties, degradation of microstructure, matrix softening

1. Introduction

One of the ways to reduce air emissions of pollutants arising from combustion of solid fuel—hard coal or lignite in power units is to enhance steam parameters. By improving the efficiency of power units, higher steam parameters contribute to the reduction in emission

Material	C	Si	Mn	P	S	Cr	Ni	W	Co	Cu	Nb	N	B
TP347HFG	0.04–0.10	max 1.00	max 2.00	max 0.04	max 0.015	17.0–20.0	9.0–12.0	–	–	–	max 1.20 (10× C)	max 0.10	–
Super304H	0.07–0.13	max 0.30	max 1.00	max 0.04	max 0.01	17.0–19.0	7.5–10.5	–	–	2.5–3.5	0.30–0.60	0.05–0.12	0.001–0.010
HR3C	0.04–0.10	max 0.75	max 2.00	max 0.03	max 0.03	24.0–26.0	17.0–23.0	–	–	–	0.20–0.60	0.15–0.35	–
Sanicro 25	0.04–0.11	max 0.40	max 0.60	max 0.025	max 0.015	21.5–23.5	23.5–26.5	2.0–4.0	1.0–2.0	2.0–3.5	0.30–0.60	0.15–0.30	max0.008

Table 1. Chemical composition of modern creep-resistant austenitic stainless steel, %mass.

http://dx.doi.org/10.5772/intechopen.70941

of pollutants. Higher requirements related to the enhancement of operating parameters of unit demands the use of new construction materials not only of adequate creep strength, but also with high resistance to corrosion and oxidation [1, 2].

Because of insufficient resistance to oxidation of 9%Cr martensitic steels and very unstable microstructure associated with the MX → Z phase transition for 12%Cr martensitic steel, in plants operated at above 600°C, creep-resistant austenitic steels are used [3]. Compared to ferritic steels, the austenitic ones show higher heat and high-temperature creep resistance, however they have unfavourable physical properties, that is, higher thermal expansion coefficient and lower conductivity. The disadvantage of these steels is high price due to the content of expensive nickel in their chemical composition. Creep-resistant austenitic steels were developed by modification and optimisation of chemical composition of the classic 18/8-grade steels [4]. Nowadays, the (17–20%)Cr-(7–11%)Ni steels—TP347HFG and Super 304H and the (20–25%)Cr-(15–23%)Ni steels—HR3C are used in modern power boilers. The prospective steel Sanicro 25, classified as steel of the 25/23 type, is also in the phase of implementation to modern power units designed for work at the ultra-supercritical parameters of steam. Austenitic steels are intended for plant components working at above 600°C, and currently they are most often used for steam superheaters [2, 4–7]. The required chemical composition of the creep-resisting austenitic steels mentioned above is presented in **Table 1**.

In the as-received state, austenitic steels have austenitic structure with numerous annealing twins. Modern creep-resistant austenitic steels belong to the so-called stabilised steels, which mean they contain a strong carbide-forming element—niobium. Therefore, numerous randomly arranged primary NbC carbides with micrometric size are observed within the matrix of these steels. Sanicro 25 steel in the as-received condition may also include the primary Z phase (complex NbCrN nitride) precipitates. By binding carbon atoms, the primary NbC carbide precipitates reduce the processes of precipitation of $M_{23}C_6$ carbides at the grain boundaries. However, the niobium content in creep-resistant austenitic steels, in contrast to the "classic" austenitic steels, is limited below the level required for complete binding of carbon atoms [5, 7]. The metastable structure of austenitic steels will be subject to progressive evolution during service, and the main degradation mechanism includes the precipitation processes of secondary phases and changes in their morphology. Depending on steel grade and operating parameters, in the microstructure of austenitic steel, there may occur during long-term service the precipitation of carbides/nitrides: $M_{23}C_6$, MX, Z phase; intermetallic phases: σ, Laves, χ, complex silicide—G phase, as well as copper-rich precipitates—ε _Cu [5, 7–12].

This paper presents the characteristics of secondary phases occurring in modern creep-resistant austenitic steels.

2. $M_{23}C_6$ carbide

In the majority of austenitic steels (without copper addition), chromium-rich $M_{23}C_6$ carbide is the first secondary precipitate that appears in the microstructure of these steels. The privileged locations of $M_{23}C_6$ carbide precipitation are grain boundaries (**Figure 1**) and, in the next place, incoherent and

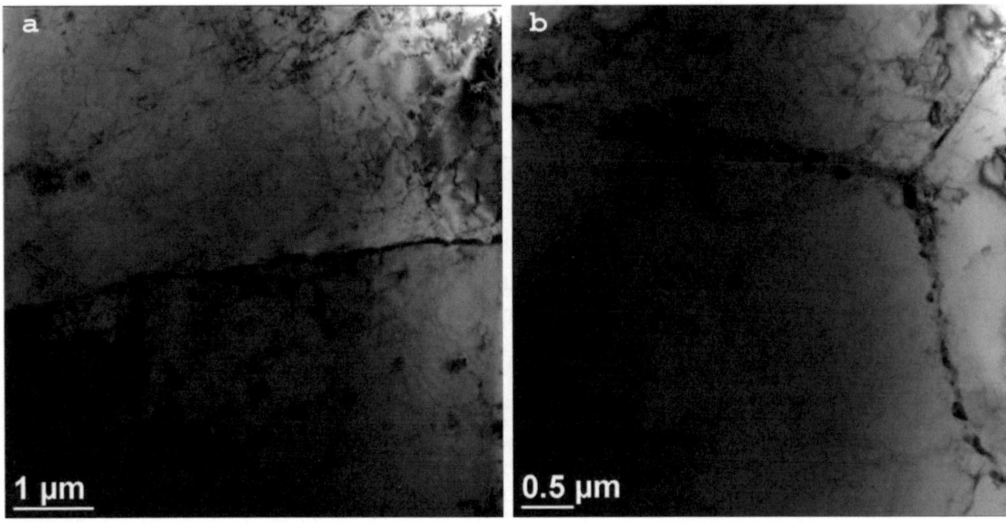

Figure 1. $M_{23}C_6$ carbide precipitates at the grain boundary in TP347HFG steel after service.

coherent twin boundaries [5, 8–10, 13]. Precipitation of $M_{23}C_6$ carbides at the grain boundaries in austenitic steels depends on the nature of the boundary. The privileged boundaries are those characterised by high degree of coincidence Σ or those with high misorientation angle Θ [9, 10].

Finely dispersed $M_{23}C_6$ carbides precipitated at the grain boundaries in the initial stage of operation hinder the slip at the grain boundaries and thus contribute to the increase in creep resistance. These carbides also inhibit the migration of grain boundaries contributing to a delay in the matrix softening process (**Figure 2**). However, $M_{23}C_6$ carbides are characterised by fairly

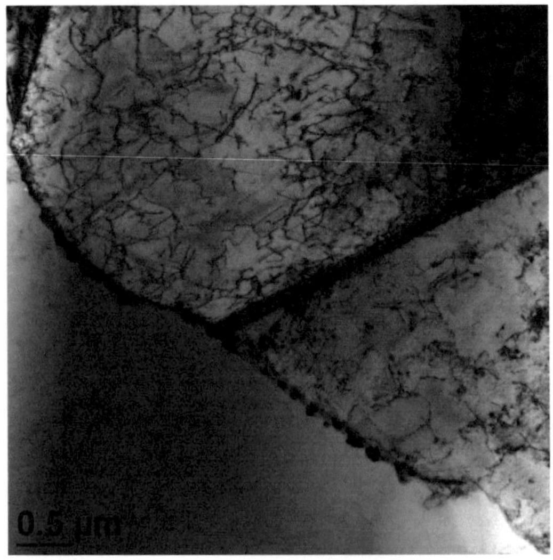

Figure 2. Impeding the grain migration by $M_{23}C_6$ carbides precipitated on the grain boundary—TP347HFG steel after service.

low thermal stability, which results in the increase in size of these precipitates (**Figure 1b**) and formation of the so-called continuous network of precipitates at the grain boundaries during service. The enthalpy of creation, which can be treated as a measure of the stability of precipitates, for the $Cr_{23}C_6$ and NbC carbides amounts to −25 and −55 kJ/mol, respectively, whereas for the NbN nitride, −125 kJ/mol [14].

The formation of the continuous network of $M_{23}C_6$ carbides at the grain boundaries and increase in their size during service results in the appearance of microareas with reduced chromium contents nearby the grain boundaries. This may result in the sensitisation of steel, that is, its increased susceptibility to intergranular corrosion. The chromium-depleted microareas constitute an anode in the corrosion process, which results in their oxidation. Such a type of corrosion is very dangerous because the process of destruction (oxidation) runs very fast throughout the material on the grain boundaries, leaving no visible traces on the steel surface. The effect of sensitization of the near-boundary areas of grains in the austenitic steel is also the disturbance in the material consistency, which can result in falling out of single grains, for example, influenced by the metallographic reagent (**Figure 3**).

Numerous precipitates at the grain boundaries have also a negative impact on ductility of austenitic steels (**Figure 4**) [9, 10, 13–17]. The precipitates of $M_{23}C_6$ carbides on the grain boundaries, for instance, in the case of HR3C steel, constitute to a very rapid decrease in ductility in a relatively short time and favour the brittle cracking with the intercrystalline mechanism (**Figure 5**) [17, 18].

The introduction of micro-addition of boron into the chemical composition of austenitic steel results in partial replacement of carbon atoms with the boron ones in $M_{23}C_6$ carbides. This causes the precipitation of $M_{23}(C, B)_6$ borocarbides in place of "pure" $M_{23}C_6$ carbides. Compared to $M_{23}C_6$ carbide, the $M_{23}(C, B)_6$ borocarbides are distinguished by higher thermal stability due to a better fit of the crystalline network of precipitates and matrix. As a consequence, the $M_{23}(C, B)_6$ borocarbides keep their finely dispersed form over a longer time during service. This results in reduction in the tendency towards the grain slip, and thus reduction in susceptibility of steel to intercrystalline cracking during creep [19, 20]. A similar favourable effect of micro-addition of boron on the increase in stability of $M_{23}C_6$ carbides was also observed in the 9–12% Cr high-chromium martensitic steels [21].

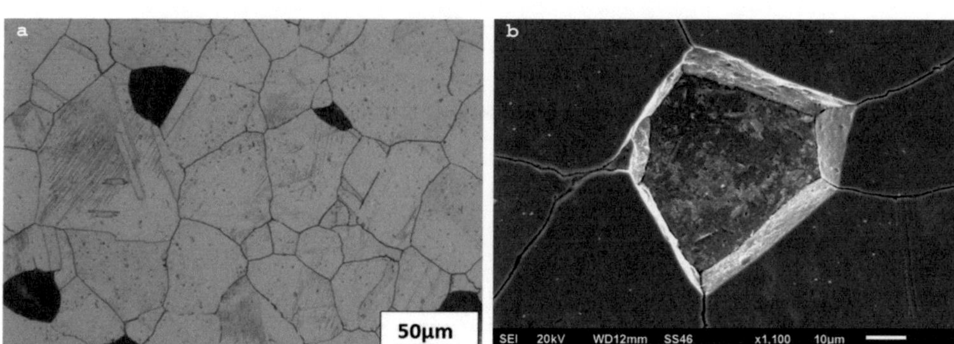

Figure 3. Austenite grains fallen out in HR3C steel after service [17].

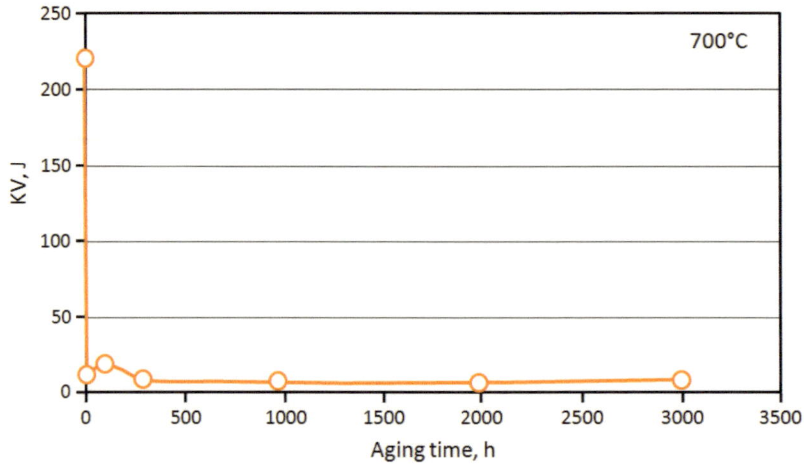

Figure 4. The impact energy changes with ageing time at 700°C in HR3C [7].

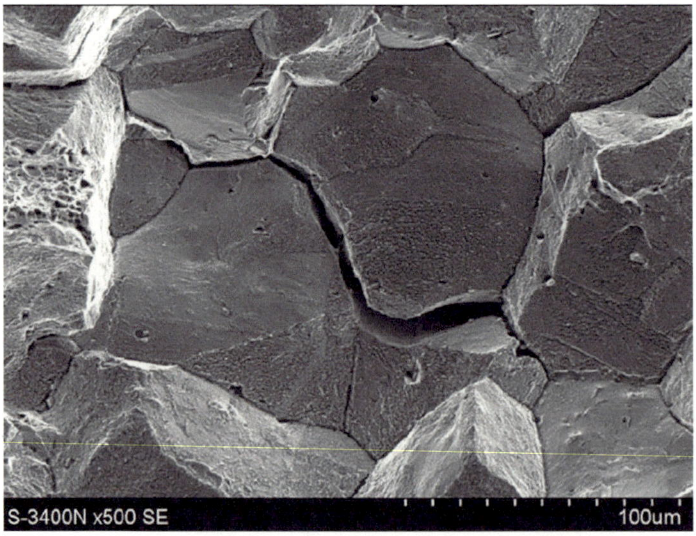

Figure 5. The intercrystalline mechanism of cracking with secondary cracks of HR3C steel after 1000 h of ageing at the temperature of 650°C [18].

3. MX precipitate

The MX secondary phases in austenitic steels are developed by carbide formers such as titanium, niobium, or hafnium. The MX precipitates in austenitic steels are ones of the most advantageous secondary phases [9, 10, 22, 23]. In austenitic steels, the MX precipitates have two main objectives:

- binding carbon atoms and preventing the precipitation of chromium-rich carbides at the grain boundaries and
- precipitation hardening of austenitic steels.

In the microstructure of niobium-stabilised austenitic steels, the presence of two types of MX (MC) precipitates is observed:

- large primary carbides of micrometric size (**Figure 6**) and
- finely dispersed secondary precipitates of nanometric size (**Figure 7**).

In addition to binding of carbon atoms, the NbC primary carbide precipitates effectively inhibit the grain growth during heat (thermomechanical) treatment. However, the NbC primary carbides should be treated as disadvantageous precipitates because the nucleation and growth of creep cracks may occur at their carbide/matrix interface.

The NbC primary carbides are characterised by very high stability: no increase in their size was observed even after 70,000 h ageing at 700°C.The NbX secondary precipitates occur during service within the grains, mainly at dislocations and stacking fault areas [9, 22, 24]. These precipitates form very effective barriers to free displacement of dislocations, pin and inhibit the possible motion of dislocations (**Figure 7**) and, in spite of their low volume fraction, have very strong effect on steel hardening. The effect of MC and $M_{23}C_6$ carbides on creep strength of the 18Cr10NiTiNb steel at 650°C can be illustrated with the following formula (1) [25]:

$$R_{z/650/10\,000} = 21.5 \text{ x } (\text{C as MC}) + 6 \text{ x } \left(\text{C as } M_{23}\, C_6\right) + 85.32 \text{ (MPa)} \tag{1}$$

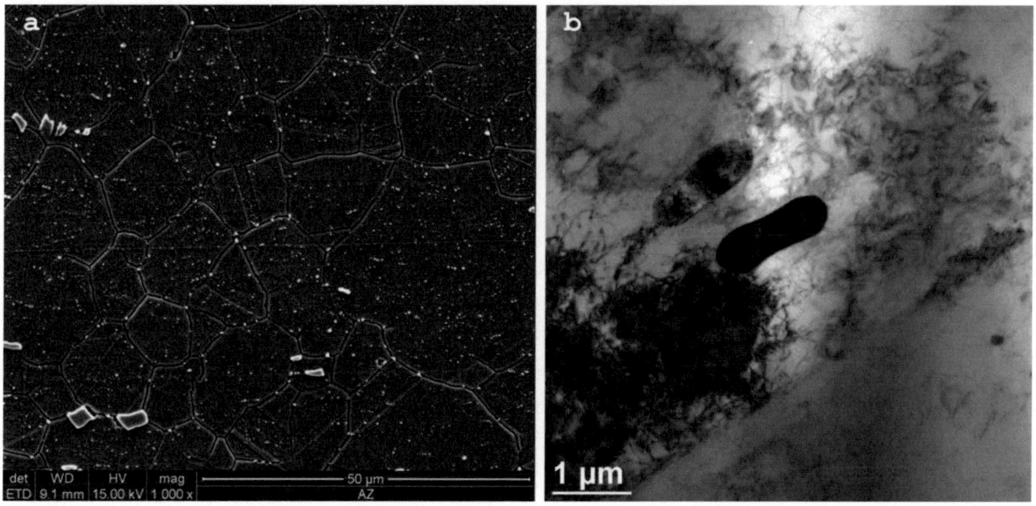

Figure 6. The NbC primary precipitates in TP347HFG steel.

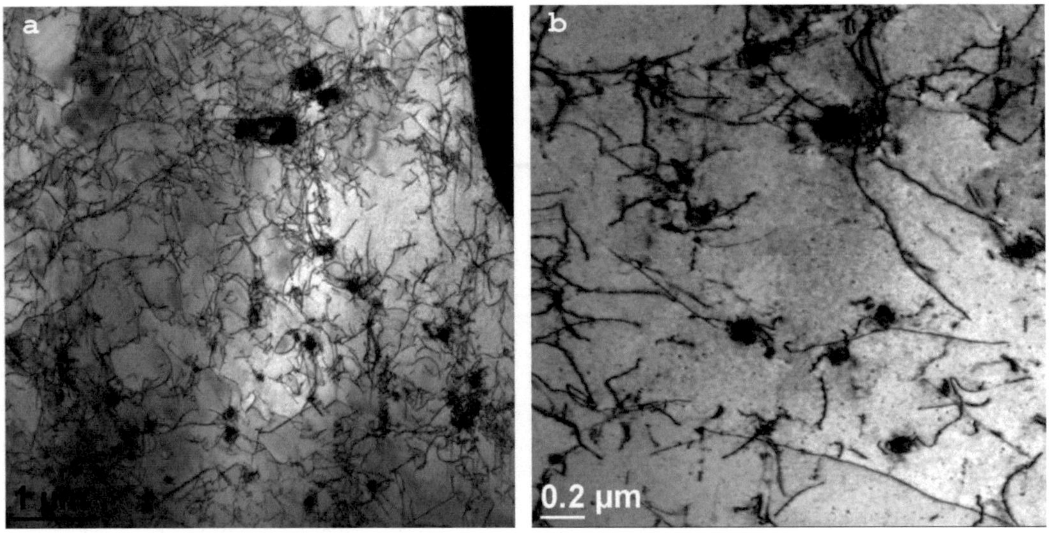

Figure 7. The NbX secondary precipitates in TP347HFG steel (a) and their interaction with dislocations (b).

The MX secondary precipitates are also characterised by high thermal stability up to approx. 700°C, which results in a very slow increase in size of these precipitates during long-term service. The MX precipitates binding the atoms of nitrogen can also cause the growth of the steel susceptibility to pitting corrosion [26].

4. ε_Cu precipitate

In austenitic steels containing a copper addition in their chemical composition, for example, Super 304H or Sanicro 25, one of the main hardening mechanisms that occur during service is the precipitation hardening with ε_Cu copper-rich particles (**Figure 8**).

The ε_Cu particles can also dissolve the atoms of iron, chromium and nickel, and their contents in a precipitate that depends on the temperature and time. These precipitates at the initial stage of service/ageing are the precipitates rich in iron, chromium and nickel, and the content of copper in the particle does not exceed 20% then (**Figure 8**). As the time passes, the precipitates get rich in copper and after around 500 h of ageing at the temperature of 650°C they constitute around 90% of copper [7, 27, 28]. Precipitation of these particles takes place very quickly. In Super 304H steel, the precipitates of this type and of approx. 2 nm were revealed only after 1 h of ageing at 650°C [7, 27, 29]. In spite of their low volume fraction (approx. 3%), finely dispersed ε_Cu precipitates (whose density in Super 304H steel after ageing at 650°C for 10,000 h was 0.38×10^4 m^{-2} and average particle diameter was 35 nm) represent a very effective barrier to free displacement of dislocations. The calculated shear stress for these coherent precipitates is approx. 37 MPa [26]. These precipitates are also characterised by quite high stability (**Figure 9**). The Cu-rich phase is growing during 650°C long time ageing till 10,000 h, but the growth rate

http://dx.doi.org/10.5772/intechopen.70941

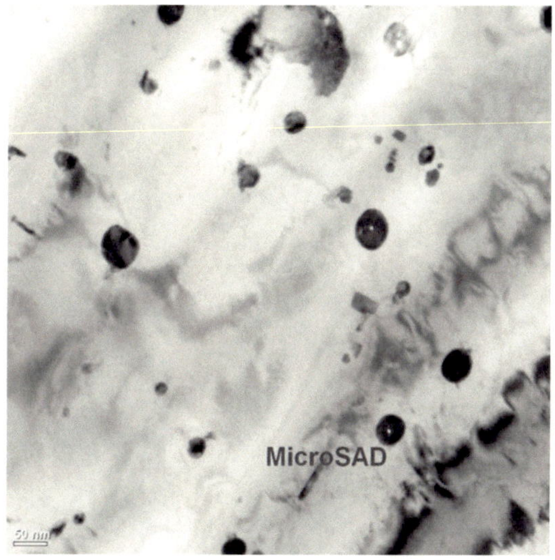

Figure 8. The ε_Cu precipitates in aged Super 304H steel.

is very slow. The average size of Cu-rich phase still keeps about 34 nm at 650°C ageing for 10,000 h [7]. This is mainly due to low energy at the ε_Cu precipitate/matrix interface, which is approx. 0.017 J/m^2, and the fact that these precipitates are coherent with the matrix.

This translates into slow growth of the copper-rich phase particles during the service and has a positive impact on the maintenance of high properties over a long time of service, especially that the volume fraction of the copper-rich precipitates grows with the time of ageing (**Figure 10**).

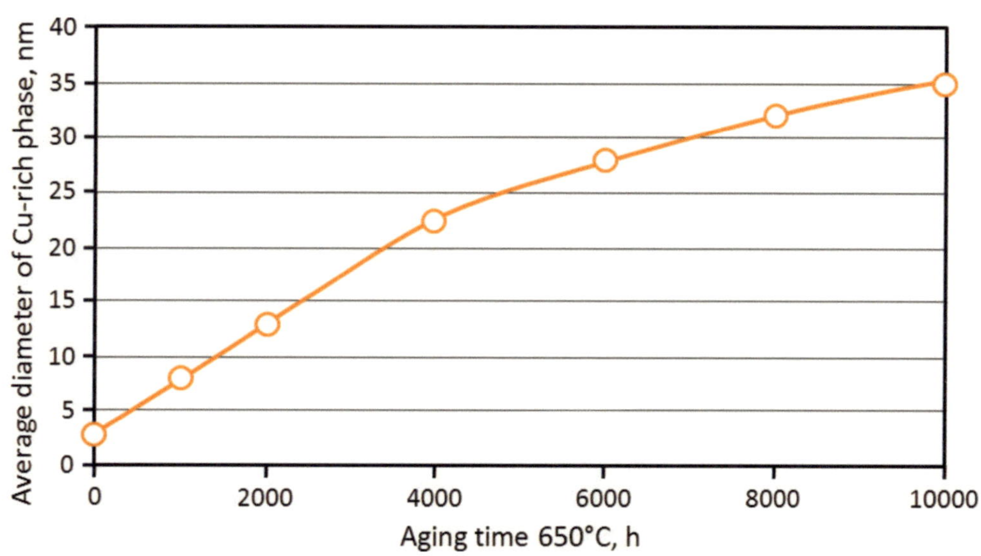

Figure 9. The effect of the time of Super304H ageing at 650°C on increase in the volume of ε_Cu precipitates [7].

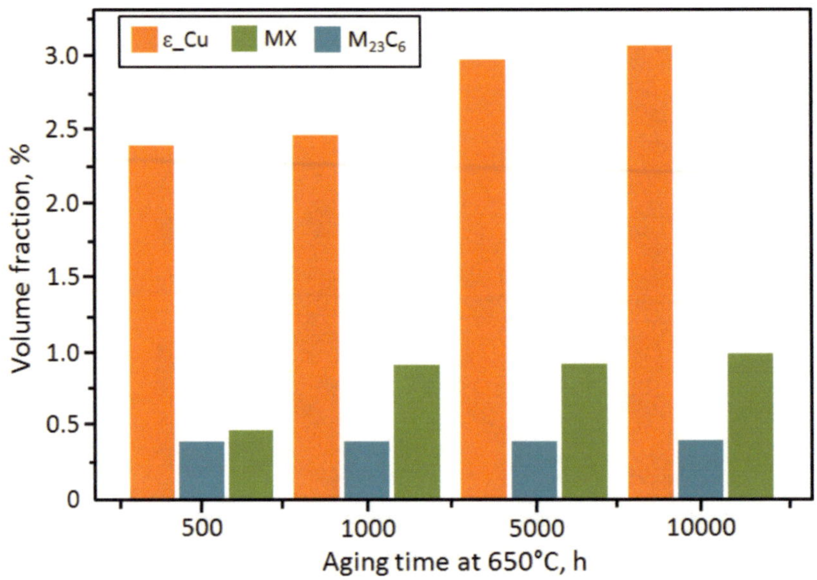

Figure 10. The precipitates volume fraction in Super304H at 650°C with ageing time [7].

However, the precipitation of ε_Cu particles has a destabilising effect on the passive layer and reduces the corrosion resistance of the steel.

5. Z phase

The NbCrN (Z phase) precipitate is privilegedly precipitated in austenitic steels with high niobium and nitrogen contents, mainly at the grain boundaries and inside the grains (**Figure 11**). In Sanicro 25 steel, both the primary and secondary Z phase precipitates are observed. In contrast to 9–12% Cr martensitic steels, the Z phase precipitated in austenitic steels has a positive impact on mechanical properties. NbCrN precipitates are characterised by high thermodynamic stability up to approx. 700°C, which results in a slow increase in the size of these precipitates. Finely dispersed NbCrN particles precipitated inside the grains cause precipitation hardening, thus contributing to the increase in creep strength [7, 9, 10, 30, 31] (**Figure 11**).

The calculated value of stress required for dislocation to bypass the precipitate with Orowan mechanism for the Z phase in Sanicro 25 steel at 700°C after 4265 h creep with average precipitate diameter of approx. 14 nm and volume fraction of 0.014 amounts to 118.7 MPa. And after 12,920 h creep at 700°C, with average diameter of Z phase secondary particles of approx. 18 nm and their volume fraction in Sanicro 25 steel amounting to 0.016, it is 105.7 MPa. For comparison, the estimated value of the stress after 4265 h creep at 700°C for ε_Cu particles in Sanicro 25 steel was 44.6 MPa [31].

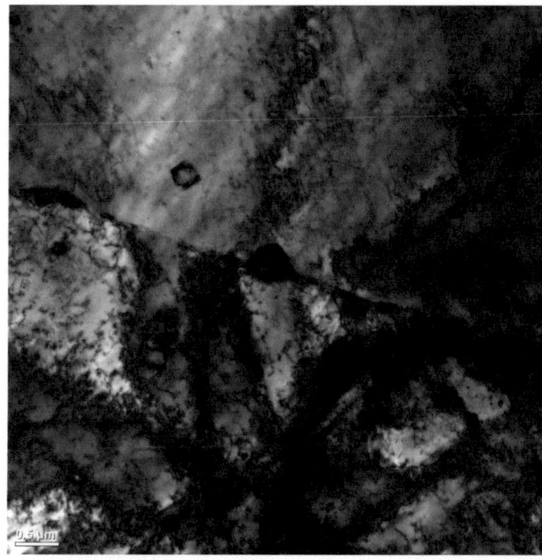

Figure 11. The Z phase precipitation at the grain boundary in Super 304H steel.

6. σ phase

The σ phase in austenitic steels is a secondary intermetallic phase which precipitates mainly at the grain boundaries, and the particularly privileged locations are the contact point of three grain boundaries and delta ferrite precipitates (**Figure 12**). The σ phase precipitation at the grain boundaries may be accompanied by dissolution of $M_{23}C_6$ carbides in the matrix

Figure 12. The σ phase precipitation at the contact point of three grain boundaries in T321H steel after long-term service.

[9, 10, 31–34]. The precipitation of σ phase in austenitic steels is much slower than that of $M_{23}C_6$ carbides, which results mainly from the factors such as (1) diffusion of substitution elements, for example, chromium, takes place slowly in austenite, (2) σ phase precipitates in austenite are incoherent with the matrix, (3) solubility of carbon and nitrogen in austenite is low, hence the privileged precipitation processes is the formation of carbides and/or nitrides [9, 11, 33–36]. In addition to chromium, other ferrite formers also create favourable conditions for σ phase precipitation. Particularly strong effect is shown by stabilising elements (carbide formers), that is, titanium and niobium. In austenitic steels with higher titanium and niobium contents, the σ phase is formed much easier due to the fact that carbon and/or nitrogen atoms are bound into MX precipitates. The rate and temperature of the σ phase precipitation in austenitic steel grow with the increase in chromium content in alloy (**Figure 13**). Also silicon shows as intensive effect on the acceleration of the σ phase precipitation process as carbide formers. The increase in the content of silicon, as an element which enhances heat resistance of austenitic steels from 0.17 to 0.76% wt, may intensify the precipitation of the σ phase and its volume fracture increases from 6 to 22%. The element that has a strong effect on delay in the σ phase precipitation is carbon, which forms $M_{23}C_6$ carbides and thus decreases chromium content in the matrix [9, 10, 32, 37].

The effect of chemical composition on susceptibility of the specific steel to σ phase precipitation is shown by the following formula (2) [39]:

$$\begin{aligned} Cr_{eq} &= Cr + 0.31Mn + 1.76Mo + 0.97W + 2.02V + 1.58Si \\ &+ 2.4Ti + 1.76Nb + 1.22Ta - 0.226Ni - 0.177Co \end{aligned} \quad (2)$$

In the event when chromium equivalent Cr_{eq} is higher by 17–18% of the weight, the steel shows a strong tendency to σ phase precipitation.

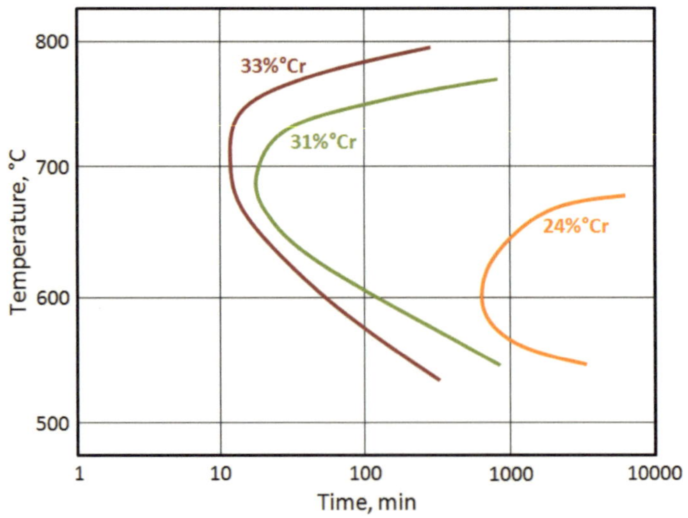

Figure 13. The effect of chromium content on the incubation period of σ phase precipitation [38].

The tendency for the σ phase to precipitate in the austenitic steel can also be shown with the dependency (3) based on the model of positive holes [33]:

$$N_V = 0.66\%Ni + 2.66\%Fe + 4.66\%(Cr + Mo) + 5.6\%Nb + 6.66\%Si + 7.66\%Al \quad (3)$$

In the case when the value N_v is bigger than the number 2.52, there is a tendency for the σ phase to precipitate in the alloy.

The σ phase precipitation in austenitic steels at the grain boundaries is a very unfavourable phenomenon as it increases the brittleness of steel and its susceptibility to sensitisation and pitting corrosion. The σ phase has also a negative effect on the plastic properties determined in static tensile test, that is, elongation and reduction in area. The effect of the σ phase on creep strength of austenitic steels is ambiguous. In the initial stage of precipitation, the finely dispersed σ phase precipitates may have, due to precipitation hardening, a positive impact on the increase in creep resistance. However, other researchers show its negative role [11, 32, 34, 37, 40].

7. Laves phase

The intermetallic Laves phase (Fe_2Nb, Fe_2Ti, Fe_2Mo) precipitates in austenitic steel at above 600°C. The privileged locations of the Laves phase precipitation are grain boundaries and, in the second place, the interior of grains. The Laves phase is a more efficient and stable precipitate than $M_{23}C_6$ carbide, it coagulates more slowly at above 600°C, and its effect on steel properties depends on the fraction of its volume at the grain boundary. Higher surface fraction of the Laves phase at the grain boundaries affects the increase in creep resistance of austenitic steels. The Laves phase inhibits dislocation slip at the grain boundaries, and thus results in the increase in creep resistance while reducing the elongation. At the same time, the Laves phase precipitates at the grain boundaries have an adverse effect on ductility of austenitic steels and their high-temperature strength properties determined in static tensile test. The finely dispersed Laves phase particles precipitated inside the grains have a positive impact on creep and fatigue strength. The effect of the Laves phase precipitated inside the grains on steel properties depends on its size and volume fraction. The coagulation of the Laves phase during service results in disappearance of this effect [9, 10, 31, 41–44].

8. G phase

The G phase is a complex silicide with general formula of $A_{16}D_6Si$ where A—nickel, D—niobium or titanium (e.g. $Ni_{16}Nb_6Si_7$; $Ni_{16}Ti_6Si_7$; $(Ni, Fe, Cr)_{16}(Nb, Ti)_6Si_7$). The G phase precipitates in titanium or niobium-stabilised austenitic steels after long-term service. Like most secondary phases precipitating in austenitic steels during service, the G phase is located at the grain boundaries (**Figure 14**) [9, 10, 32, 44, 45]. The G phase is formed in austenitic steels due to the *in situ* transition of primary NbC carbides and/or enrichment of $M_{23}C_6$ carbides in silicon. The *in situ* NbC carbide → G phase transition takes place as a result of enrichment due to NbC diffusion into

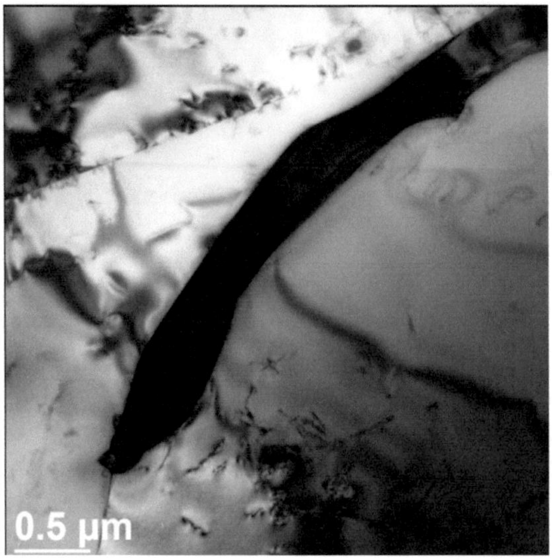

Figure 14. The G phase precipitation at the grain boundaries in T321H steel after long-term service.

nickel and silicon atoms. The rate of transition—G phase formation depends strongly on silicon content in steel. Higher silicon contents result in reduction in the time of G phase incubation [9, 15].

The effect of the G phase on properties of austenitic steels is ambiguous. The G phase precipitated at the grain boundaries has a positive impact on elongation in creep test and delays the secondary recrystallisation processes by inhibiting the migration of grain boundaries [9, 10, 15, 44–47].

The finely dispersed G phase precipitates at the grain boundaries have also a positive impact on creep resistance of austenitic steels during service at around 750°C. Nevertheless, like any precipitate at the grain boundary, the G phase has also a negative impact on steel ductility. The scale of this negative impact is affected by the amount and size of precipitates at the boundary. The formation of the G phase is accompanied by matrix depletion of nickel, which affects the increase in instability of the matrix and has a negative impact on corrosion resistance. The matrix depletion of nickel also results in the increase in susceptibility to σ phase precipitation in these steels [9, 10, 15, 45–47].

9. χ phase

The intermetallic χ phase precipitates mainly in high-nitrogen steels with addition of molybdenum. Similarly to $M_{23}C_6$ carbides, the privileged locations of the χ phase precipitation are primarily the grain boundaries, and further the incoherent and coherent twin boundaries as well as the dislocations within the grains. In the initial stage of precipitation and growth at the grain boundaries, the dispersive χ phase has a positive impact on creep resistance of austenitic steels as these precipitates hinder slip at the grain boundaries. The increase in size of the χ phase precipitated at the grain boundaries during service has an adverse impact on ductility, which is related to reduction in impact strength. The χ phase precipitated at the grain boundaries in austenitic steel increases the susceptibility of the alloy to intercrystalline cracking. These precipitates also result in steel sensitisation. The χ phase precipitates at the boundaries have a negative impact on creep strength of the steel too. The individual effect of the χ phase on performance properties of austenitic steels is difficult to define, because this phase occurs together with the σ phase in the microstructure. In austenitic steels operating at above 700°C, the χ phase may be transformed into the intermetallic σ phase according to the following diagram [9, 10, 32, 48].

The effect of the secondary phases precipitating in creep-resistant austenitic steels during service on their performance properties is summarised in **Table 2**.

Phase	Location	Positive effect	Negative effect
$M_{23}C_6$ carbides	Grain boundaries, incoherent and coherent twin boundaries, inside the grain	• Finely dispersed carbides precipitated at the grain boundaries increase creep resistance; • Inhibition of secondary recrystallisation	✓ Increase in embrittlement ✓ Sensitisation of steel
MX precipitates	Inside the grains at dislocations	• Precipitation hardening • Increase in hardness and strength properties • Increase in creep strength	✓ At the MX primary precipitate/matrix interface, there may occur the nucleation and growth of creep cracking
Z phase (NbCrN)	Inside the grains at dislocations	• Precipitation hardening • Increase in hardness and strength properties • Increase in creep strength	✓ Disappearance of MX precipitates
σ phase	Grain boundaries, incoherent twin boundaries, inside the grain	–	✓ Increase in embrittlement ✓ Sensitisation of steel ✓ Deteriorates strength and plastic properties ✓ Reduces temporary creep strength
G phase	Grain boundaries	• Increases the value of elongation at creep • Delays secondary recrystallisation	✓ Reduces temporary creep strength ✓ Results in decrease in corrosion resistance
Laves phase	Grain boundaries	• Increases creep resistance • Increases fatigue strength	✓ Reduces elongation ✓ Deteriorates ductility ✓ Deteriorates high-temperature strength properties
X phase	Grain boundaries, incoherent twin boundaries, inside the grain	• Hinders slip at the grain boundaries—increases creep strength	✓ Sensitisation of steel ✓ Increase in embrittlement
ε_Cu	Inside the grain	• Precipitation hardening • Increase in strength properties and creep resistance	✓ Reduction in corrosion resistance

Table 2. The effect of secondary phase precipitates on properties of austenitic steels.

10. Summary

The paper describes the influence of particular precipitates on the microstructure and functional properties of creep-resisting austenitic steels. The structure of austenitic steels in the supersaturated state is a non-equilibrium structure which will undergo gradual degradation during the service at elevated temperature, which will further influence the decrease in mechanical properties and corrosion resistance/ oxidation resistance of these materials. The precipitation processes running during the service of austenitic steels are the basic mechanism having a very significant effect on the functional properties of these steels. The type of precipitates and their volume fraction depends on the chemical composition of the given steel, the solubility limit of carbon and alloy elements in the matrix, and the service history (temperature, time, pressure, number and sort of shutdowns, etc.). The main factor influencing the precipitation processes for a given sequence of precipitation of secondary phases during the service is the temperature of work. Precipitations in austenitic steels can be divided in terms of their influence on the microstructure and functional properties into the favourable ones, including nitrides/carbonitrides NbX, Z phase and εCu precipitates, and the harmful ones, including σ, χ and G phases. However, the influence of $M_{23}C_6$ carbides and Laves phase on the properties of austenitic steels depends on their morphology.

Particularly dangerous precipitates in austenitic steels are the ones precipitating on the grain boundaries, σ phase and $M_{23}C_6$ carbides. Therefore, the growth of stability of $M_{23}C_6$ carbides and the delay or even inhibition of precipitation of the σ phase, or using the strengthening with intermetalic phases seem to be the main directions of further development of the creep-resisting austenitic steels.

Author details

Grzegorz Golański[1]*, Adam Zieliński[2] and Hanna Purzyńska[2]

*Address all correspondence to: grisza@wip.pcz.pl

1 Institute of Materials Engineering, Czestochowa University of Technology, Czestochowa, Poland

2 Institute for Ferrous Metallurgy, Gliwice, Poland

References

[1] Viswanathan R, Henry JF, Tanzosh JM, Stanko G, Shingledecker J, Vitalis B, Purget R. US program on materials technology for ultra-supercritical coal power plants. Journal of Materials Engineering and Performance. 2005;**14**:281-292. DOI: doi.org/10.1361/10599490524039

[2] Di Gianfrancesco A. The fossil fuel power plants technology. In: Di Gianfrancesco A, editor. Materials for Ultra-supercritical and Advanced Ultra-supercritical Power Plants. Cambridge: Woodhead Publishing; 2017. p. 1-49

[3] Danielsen HK, Hald J. Influence of Z-phase on long-term creep stability of martensitic 9 to 12%Cr steels. VGB PowerTech. 2009;**5**:68-73

[4] Mayer K–H, Masuyama F. The development of creep-resistant steels. In: Abe F, Kern T-U, Viswanathan R, editors. Creep Resistant Steels. Cambridge: Woodhead and Maney Publishing; 2008. p. 15-77

[5] Zieliński A, Dobrzański J, Purzyńska H, Golański G. Properties, structure and creep resistance of austenitic steel Super304H. Materials Testing. 2015;**57**:859-865. DOI: doi.org/10.3139/120.110791

[6] Plaut RL, Herrera C, Escriba DM, Rios PR, Padilha AF. A short review on wrought austenitic stainless steels at high temperatures: processing, microstructure, properties and performance. Materials Research. 2007;**10**:453-460. DOI: dx.doi.org/10.1590/S1516-14392007000400021

[7] Chi Ch YH, Xie X. Advanced austenitic heat-resistant steels for ultra-super-critical (USC) fossil power plants. In: Morales EV, editor. Alloy steel—Properties and Use. Rijeka: InTech Publ; 2011. p. 171-201

[8] Golański G, Lis AK, Słania J, Zieliński A. Microstructural aspect of long term service of the austenitic TP347HFG stainless steel. Archives of Metallurgy and Materials. 2015;**60**:2091-2094

[9] Sourmail T. Precipitation in creep resistant austenitic stainless steel. Materials Science and Technology 2001;**14**:1-14. DOI: dx.doi.org/10.1179/026708301101508972

[10] Padilha AF, Rios PR. Decomposition of austenite in austenitic stainless steels. ISIJ International. 2002;**42**:325-337. DOI: doi.org/10.2355/isijinternational.42.325

[11] Lai JKL. A review of precipitation behaviour in AISI Type 316 stainless steel. Materials Science and Engineering. 1983;**61**:101-109. DOI: doi.org/10.1016/0025-5416(83)90191-X

[12] Ayer R, Klein CF, Marzinsky CN. Instabilities in stabilized austenitic stainless steels. Metallurgical Transactions. 1992;**23A**:2455-2467. DOI: 10.1007/BF02658049

[13] Kaneko K, Futunaga T, Yamada K, Nakada N, Kikuchi M, Saghi Z, Barnad JS, Midgley PA. Formation of $M_{23}C_6$-type precipitates and chromium-depleted zone in austenite stainless steel. Scripta Materialia. 2011;**65**:509-512. DOI: doi.org/10.1016/j.scriptamat.2011.06.010

[14] Pickering FB. Historical development and microstructure of high chromium ferritic steels for high temperature applications. In: Strang A, Grooch DJ, editors. Microstructural Development and Stability in High Chromium Ferritic Power Plant Steel. Cambridge: The Institute of Materials; 1997. p. 1-20

[15] Wang B, Liu D, S-Ch C, Ch-M L, Wang JZ. Microstructure evolution and mechanical properties of HR3C steel during long-term aging at high temperature. Journal of Iron and Steel Research, International. 2014;**21**:765-773. DOI: doi.org/10.1016/S1006-706X(14)60139-4

[16] Saucedo-Munoz ML, Lopez-Hirata VM, Avila-Davila EO, Melo-Maximo DV. Evaluation of toughness deteriotation by an electrochemical method in an isothermally-aged N-containing austenitic stainless steel. Materials Characterization. 2009;**60**:119-124. DOI: doi.org/10.1016/j.matchar.2008.07.007

[17] Golański G, Kolan C, Zieliński A, Klimaszewska K, Merda A, Sroka M, Kłosowicz J. Microstructure and mechanical properties of HR3C steel after service. Archives of Materials Science and Engineering. 2016;**81**:62-67

[18] Zieliński A, Sroka M, Hernas A, Kremzer M. The effect of long-term impact of elevated temperature on changes in microstructure nad mechanical propertis of HR3C steel. Archives of Metallurgy and Materials. 2016;**61**:761-766. DOI: doi.org/10.1515/amm-2016-0129

[19] Laha K, Kyono J, Sasaki T, Kishimoto S, Shinya N. Improved creep strength and creep ductility of type 347 austenitic stainless steel through the self-healing effect of boron for creep cavitation. Metallurgical and Materials Transactions A. 2005;**36A**:399-409. DOI: 10.1007/s11661-005-0311-0

[20] Laha K, Kyono J, Kishimoto S, Shinya N. Beneficial effect of B segregation on creep cavitation in a type 347 austenitic stainless steel. Scripta Materialia. 2005;**52**:675-678. DOI: doi.org/10.1016/j.scriptamat.2004.11.016

[21] Abe F. New martensitic steels. In: Di Gianfrancesco A, editor. Materials for Ultra-Supercritical and Advanced Ultra-Supercritical Power Plants. Cambridge: Woodhead Publishing; 2017. p. 323-374

[22] Erneman J, Schwind M, Andrén H-O, Nilsson J-O, Wilson A, Ǻgren J. The evolution of primary and secondary niobium carbonitrides in AISI 347 stainless steel during manufacturing and long-term ageing. Acta Materialia. 2006;**54**:67-76. DOI: doi.org/10.1016/j.actamat.2005.08.028

[23] Solenthaler C, Ramesh M, Uggowitzer PJ, Spolenak R. Precipitation strengthening of Nb-stabilized TP347 austenitic steels by a dispersion of secondary Nb(C,N) formed upon a short-term hardening heat treatment. Materials Science and Engineering: A. 2015;**647A**:294-302. DOI: doi.org/10.1016/j.msea.2015.09.028

[24] Zhou Y-H, Ch-X L, Liu Y-C, Guo Q-Y, Li H-J. Coarsening behavior of MX carbonitrides in type 347H heat-resistant austenitic steel during thermal aging. International Journal of Minerals, Metallurgy, and Materials. 2016;**23**:283-293. DOI: 10.1007/s12613-016-1237-9

[25] Minami Y, Kimura H. Effect of $M_{23}C_6$ and MC carbides on the creep rupture strength of 18%Cr-10%Ni-Ti-Nb steel. Transactions of the Iron and Steel Institute of Japan. 1989;**27**:299-301. DOI: 10.2355/isijinternational1966.27.299

[26] Bai G, Lu S, Li D, Li Y. Influence of niobium and solution treatment temperature on pitting corrosion behaviour of stabilised austenitic stainless steel. Corrosion Science. 2016;**105**:111-124. DOI: doi.org/10.1016/j.corsci.2016.03.009

[27] Ou P, Xing H, Wang XL, Sun J. Tensile yield behavior and precipitation strengthening mechanism in Super304H steel. Materials Science and Engineering: A. 2014;**600A**:171-175. DOI: doi.org/10.1016/j.msea.2014.01.085

[28] Chi C-Y, Yu H-Y, Dong J-X, Xie X-S, Chen X-F, Lin F. Strengthening effect of Cu-rich phase precipitation in 18Cr9Ni3CuNbN austenitic heat-resisting steel. Acta Metallurgica Sinica (English Letters). 2011;**24**:141-147

[29] Yu H, Chi C. Precipitation behavior of Cu-rich phase in 18Cr9Ni3CuNbN austenitic heat-resistant steel at early aging state. Chinese Journal of Materials Research. 2015;**29**:195-200

[30] Yang Y, Zhu L, Wang Q, Zhu C. Microstructural evolution and the effect on hardness and plasticity of S31042 heat-resistant steel during creep. Materials Science and Engineering: A. 2014;**608A**:164-173. DOI: doi.org/10.1016/j.msea.2014.04.073

[31] Zhou R, Zhu L, Liu Y, Lu Z, Chen L, Ma H. Microstructural evolution and the effect on hardeness of Sanicro 25 welded joint base metal after creep at 973K. Journal of Materials Science. 2017;**52**:6161-6172. DOI: 10.1007/s10853-017-0758-6

[32] Lo HK, Shek CH, Lai JKL. Recent developments in stainless steels. Materials Science and Engineering: A. 2009;**65R**:39-104. DOI: doi.org/10.1016/j.mser.2009.03.001

[33] Hsieh Ch-Ch WW. Overview of intermetallic sigma (σ) phase precipitation in stainless steels. International Scholary Research Network732471. 2012:1-16

[34] West D, Hulance J, Higginson RL, Wilcox GD. σ-phase precipitation in 347HFG stainless steel. Materials Science Technology. 2013;**29**:835-842. DOI: dx.doi.org/10.1179/1743284712Y.0000000100

[35] Padilha AF, Escriba DM, Materna-Morris E, Rieth M, Klimenkov M. Precipitation in AISI 316(N) during creep tests at 550 and 600°C up to 10 years. Journal of Nuclear Materials. 2007;**362**:132-138. DOI: doi.org/10.1016/j.jnucmat.2006.12.027

[36] Shwind M, Källqvist J, Nilsson J-O, Ågren J, Andrén H-O. σ-Phase precipitation in stabilized austenitic stainless steels. Acta Materialia. 2000;**48**:2473-2481. DOI: doi.org/10.1016/S1359-6454(00)00069-0

[37] Lin D-Y, TCh C, Liu GL. Effect of Si contents of the growth behavior of σ phase in SUS 309L stainless steels. Scripta Materialia. 2003;**43**:855-860. DOI: doi.org/10.1016/S1359-6462(03)00481-0

[38] Blicharski M. Materials Technology. Steel. Warsaw: WNT; 2013 (in Polish)

[39] Barnard P. Austenitic steel grades for boilers in ultra-supercritical power plants. In: Di Gianfrancesco A, editor. Materials for ultra-supercritical and advanced ultra-supercritical power plants. Cambridge: Woodhead Publishing; 2017. p. 99-119

[40] Strutt AJ, Vecchio KS. Simultaneous oxidation and sigma-phase formation in a stainless steel. Metallurgical and Materials Transactions A. 1999;**30A**:355-362. DOI: 10.1007/s11661-999-0324-1

[41] Kestenbach H-J, Bueno LO. Effect of Fe_2Nb precipitation on the creep properties of niobium-bearing austenitic stainless steels. Materials Science and Engineering: A. 1984;**66**: 19-23. DOI: doi.org/10.1016/0025-5416(84)90157-5

[42] Chen SW, Zhang C, Xia ZX, Ishikawa H, Yang ZG. Precipitation behavior of Fe_2Nb Laves phase on grain boundaries in austenitic heat resistant steels. Materials Science and Engineering: A. 2014;**615A**:183-188. DOI: doi.org/10.1016/j.msea.2014.07.104

[43] Spiegel M, Schraven P. New austenitic steels for the advanced USC power plants. In: Di Gianfrancesco A, editor. Materials for Ultra-Supercritical and Advanced Ultra-Supercritical Power Plants. Cambridge: Woodhead Publishing; 2017. p. 375-390

[44] Powell DJ, Pilington R, Miller DA. The precipitation characteristics of 20% Cr/25% Ni-Nb stabilised stainless steel. Acta Metallurgica. 1988;**36**:713-724. DOI: doi.org/10.1016/0001-6160(88)90105-8

[45] Ibañez RAP, de Almeida Soares GD, de Ameila LH, May IL. Effects of Si content on the microstructure of modified-HP austenitic steels. Materials Characterization 1993;**30**:243-249. DOI: doi.org/10.1016/1044-5803(93)90071-3

[46] Ecob RC, Lobb RC, Kohler VL. The formation of G-phase in 20/25 Nb stainless steel AGR fuel cladding alloy and its effect on creep properties. Journal of Materials Science. 1987;**22**:2867-2880. DOI: doi.org/10.1007/BF01086484

[47] Wang J-Z, Liu Z-D, Bao H-S, Cheng SC. Evolution of precipitates of S31042 heat resistant steel during 700°C ageing. Journal of Iron and Steel Research, International. 2013;**20**:113-121. DOI: doi.org/10.1016/S1006-706X(13)60186-7

[48] Anburaj J, Mohamed Nazirudeen SS, Narayanan R, Anandavel B, Chandrasekar A. Ageing of forged superaustenitic stainless steel: precipitate phases and mechanical properties. Materials Science and Engineering: A. 2012;**535A**:99-107. DOI: doi.org/10.1016/j.msea.2011.12.048

Chapter 6

Effect of Precipitation on Cryogenic Toughness of N-Containing Austenitic Stainless Steels After Aging

Maribel L. Saucedo-Muñoz and
Victor M. Lopez-Hirata

Additional information is available at the end of the chapter

http://dx.doi.org/10.5772/intechopen.70367

Abstract

This chapter shows the effect of intergranular precipitation on the cryogenic toughness of N-containing austenitic stainless steels in comparison to that for 316-type austenitic stainless steels. First part of the chapter deals with the thermodynamic stability and growth kinetics of the precipitated phases in the austenite matrix based on Thermo-Calc software. To continue, the experimental evolution of precipitation for N-containing steels is compared to that of 316-type steel and the difference between them are explained based on the Thermo-Calc PRISMA-calculated results. Finally, the effect of intergranular precipitation on the cryogenic fracture toughness is also analyzed using Charpy V-Notch impact test results. The fracture mode is also related to the precipitation characteristics.

Keywords: N-containing austenitic stainless steel, precipitation, aging, toughness

1. Introduction

Austenitic stainless steels are used to construct different equipments with good corrosion resistance in most of the principal industries, for instance, the chemical, petroleum, and nuclear power industries. These structural materials are iron alloys, which contain a minimum of about 11% chromium. This content of chromium enables the formation of a passive film, which is self-protecting in different environments [1].

Nowadays, there are more than 200 different alloys, which can be recognized as belonging to the stainless steel group, and each year new ones and modifications of existing ones appear. In some stainless steels, the chromium content now approaches 30%, and many other elements are added to provide specific properties or ease of fabrication. For example, nickel, nitrogen, and molybdenum are added for corrosion resistance; carbon, molybdenum, nitrogen, titanium, aluminum, and copper for strength; sulfur and selenium for machinability; and nickel for formability and toughness.

Nitrogen is an alloying element, which has been used in iron-based alloys since the beginning of last century and it has been studied during the last three decades. However, nitrogen steels are so far not widely employed. The reason for this is related to the old customer skepticism in relation to nitrogen is an element, which causes brittleness in the steels.

In the case of austenitic stainless steels, the main motivating force in the development of nitrogen-containing steels is the much higher yield and tensile strengths reached, in comparison with the conventionally processed austenitic stainless steels without losing its toughness. Nitrogen-containing stainless steels present yield and tensile 200–350% higher than those of the AISI 300 series steels. Moreover, the yield strength can be increased above 2 GPa by cold deformation process because of its work hardening potential. It is also important to note that, in contrast to carbon-containing austenitic steels, nitrogen-containing austenitic stainless steels keep a high fracture toughness at low temperatures, higher than 200 Pa $m^{-1/2}$ [2].

The higher mechanical properties of nitrogen-containing austenitic stainless steels have made its application very attractive in the power-generation industry, shipbuilding, railways, cryogenic process, chemical equipment, pressure vessels, and nuclear industries.

This kind of steels contains supersaturated nitrogen levels, and it is susceptible to intergranular precipitation of chromium nitride (Cr_2N) as same as the AISI 300 series stainless steels are prone to the precipitation of chromium carbide ($Cr_{23}C_6$). If the precipitation of chromium took place, the stainless steel is susceptible to sensitization, which may cause intergranular corrosion and intergranular stress corrosion cracking [2]. Nevertheless, the chromium depletion due to the nitride precipitation for nitrogen-containing steels has been reported [3, 4] to be lower than that of carbide precipitation for the AISI 300 series steels, which causes a lower degree of sensitization in the former steel.

In addition, the intergranular precipitation and coarsening of chromium-rich precipitates have been observed to cause a decrease in the toughness of austenitic stainless steels. This type of microstructural deterioration has been named as "thermal aging". It has been reported that the precipitation is very complex, being able to involve more than 20 different precipitated phases [5]. This thermal aging may also occur in the heat-affected zones during the welding of these steels causing the decrease in toughness especially at cryogenic temperatures.

Therefore, the purpose of this chapter is to describe the kinetic behavior of precipitation during the isothermal aging of N-containing austenitic steels and its effect on cryogenic toughness, as well as its comparison with 300 series austenitic stainless steels.

2. Thermodynamics and kinetics of precipitation

This section shows the thermodynamic stability of solid phases in three austenitic stainless steels whose chemical composition is shown in **Table 1**. The first chemical composition corresponds to an AISI 316L steel and the other two, designated as A1 and A2 steels, correspond to N-containing steels. All these steels can be used for cryogenic applications since they possess good fracture toughness at temperatures as low as liquid nitrogen temperature. The highest interstitial solute contents, Ni and Cr contents are for the A2 steel. A1 steel has the highest content of Mn and the lowest contents of Cr and Ni.

The thermodynamic analysis of these steels was carried out using the commercial software, Thermo-Calc [6]. **Figure 1(a–c)** shows the equilibrium pseudobinary Fe-C at temperature lower than 1000°C and low carbon contents for each of the three steels. In the case of 316 steels, **Figure 1(a)**, a solvus line is observed, which separates the austenite region, FCC_A1, from the austenite and $M_{23}C_6$ carbide, FCC_A1 + $M_{23}C_6$. $M_{23}C_6$ carbide is the more general designation of $Cr_{23}C_6$, which is the main carbide precipitated in austenitic Cr-Ni steels. It has a fcc crystalline structure. As temperature decreases, several phases are formed such as σ (SIGMA) phase, χ (CHI) phase, and ferrite (BCC_A2). σ phase has a tetragonal unit cell with a composition FeCr. χ (CHI) phase is a bcc intermetallic phase with a typical composition of $Fe_{36}Cr_{12}Mo_{10}$ [1, 2]. In the case of A1 steel, **Figure 1(b)**, there is no phase region corresponding only to the austenite phase, FCC_A1, but a phase mixture of austenite and M_2N, HCP_A#2, which usually substitute $M_{23}C_6$ in N-containing austenitic stainless steel, and it has a close packed hexagonal unit cell. This nitride is also designated as Cr_2N. As the temperature decreases, $M_{23}C_6$ becomes stable. A further decrease in temperature promotes additionally the formation of the σ phase. The A2 steel has a similar behavior for phase formation than that showed in the diagram of the A1 steel, **Figure 1(c)**.

On the other hand, **Figure 2(a–c)** shows the expected phases during aging at temperatures of 500–1000°C for these three steels. In the case of 316 steel, the precipitation of σ phase and $M_{23}C_6$ is favored in the austenite matrix. In contrast for the N-containing steels, the precipitation of M_2N, HCP_A#2, and $M_{23}C_6$ in the austenite matrix is more likely to occur during aging at these temperatures. Besides, the precipitation of the σ phase is also possible for these two steels. A remarkable point of these diagrams is that the bcc ferrite (BCC_A2) phase is clearly present in the 316 and A2 steels, but it is absent in the case of A1 steel. This is attributable to the high content of Mn, an austenite-former. This fact suggests the possible formation of delta ferrite during the solidification of 316 and A2 steels, which is associated with a brittle behavior [1]. The precipitation behavior for the steels is shown in the Time-Temperature-Precipitation (TTP) diagram for these steels, as shown in **Figure 3(a–c)**. These diagrams were calculated using Thermo-Calc PRISMA [7] for grain boundary precipitation considering the following parameters: grain size of 100 μm, interfacial energy for the interfaces $\gamma/M_{23}C_6$, $\gamma//M_2N$, and γ/σ of 0.15, 0.22, and 0.28 J/m^2, respectively [8]. Aging process was analyzed for times up to

Steel	C	N	Mn	Si	Cr	Ni	Mo
316	0.03	–	1.4	0.35	16.8	10.4	2.0
A1	0.05	0.25	21.3	0.4	13.0	9.2	1.0
A2	0.04	0.32	3.9	1.0	24.3	15.1	–

Table 1. Chemical composition of steels in wt.%.

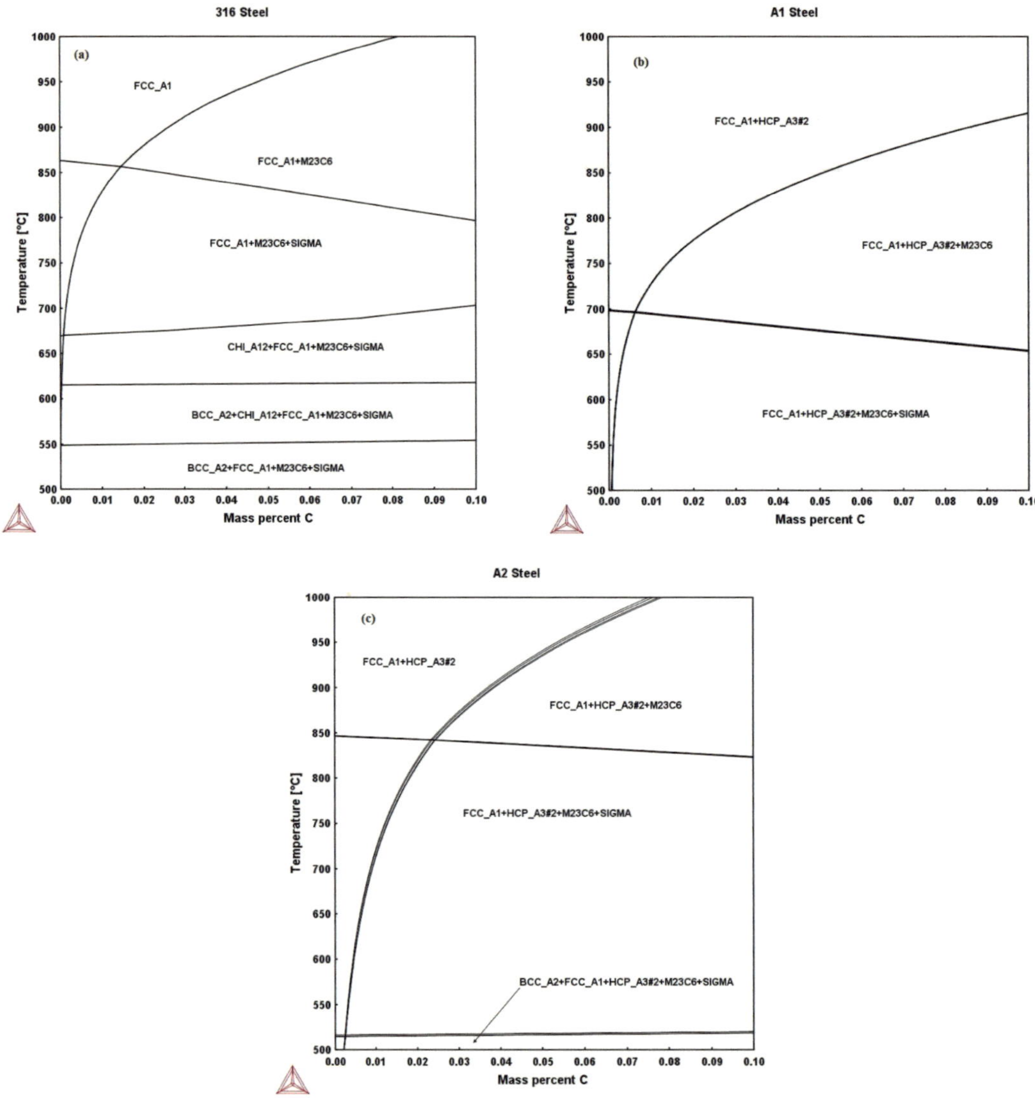

Figure 1. Equilibrium pseudobinary Fe-C phase diagrams for (a) 316, (b) A1, and (c) A2 steels.

1000 min. The TTP diagram for 316 steel is shown in **Figure 3(a)** and it can be noticed that the following precipitation reaction takes place:

$$\gamma_{sss} \rightarrow \gamma + M_{23}C_6 \tag{1}$$

The fastest growth kinetics of precipitation is at a temperature slightly lower than 700°C. Besides, no σ phase is formed at these temperatures for short aging times. This fact is in good agreement with the experimental TTP diagrams reported in the literature [1].

http://dx.doi.org/10.5772/intechopen.70367

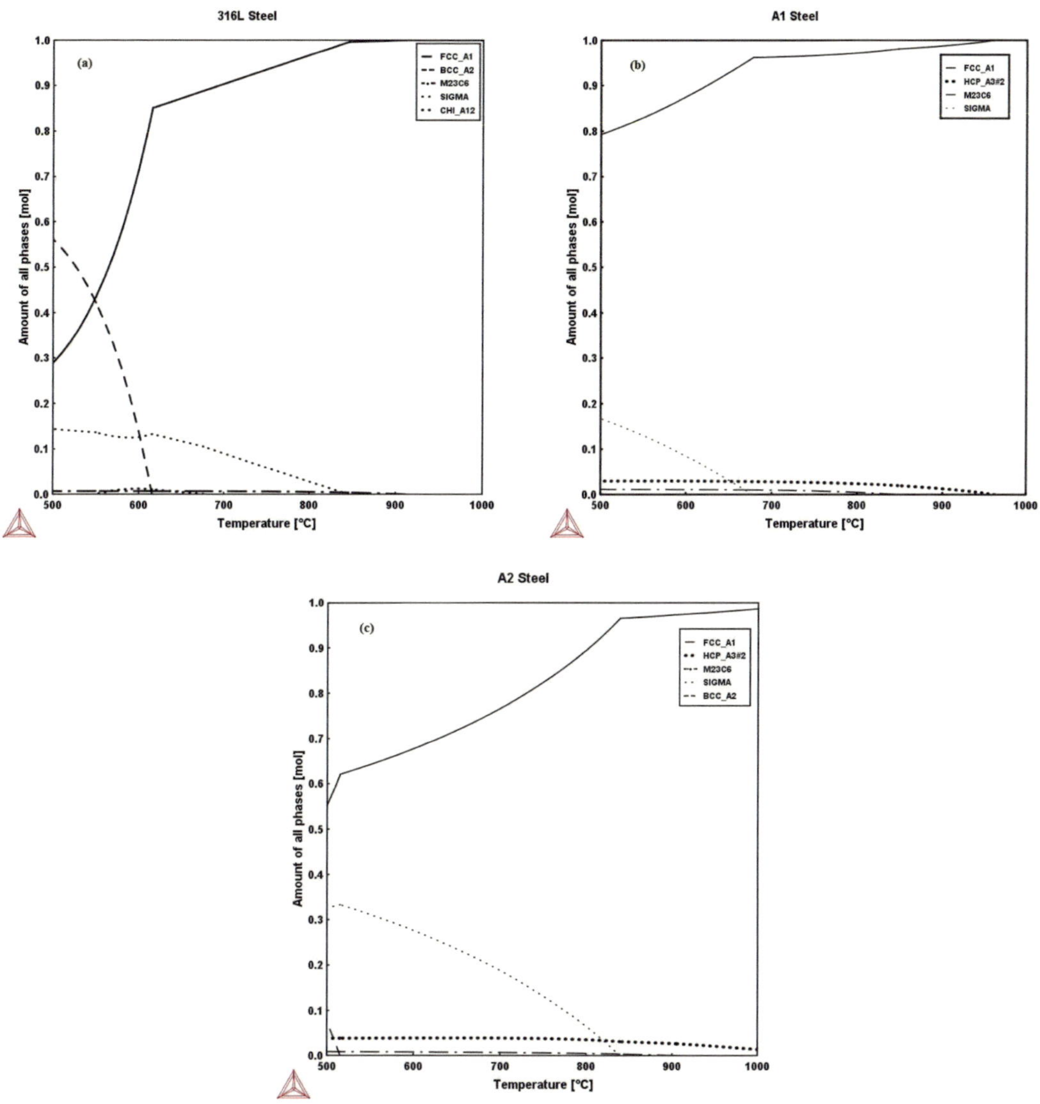

Figure 2. Amount of all phase vs. temperature for (a) 316, (b) A1, and (c) A2 steels.

In the case of both N-containing A1 and A2 steels, the following precipitation reaction is observed to occur in the TTP diagrams, **Figure 3(b, c)**:

$$\gamma_{sss} \rightarrow \gamma + M_2N \tag{2}$$

That is, the aging at these temperatures for times up to 1000 min only promotes the precipitation of nitrides. The growth kinetics of precipitation occurs more rapidly in the

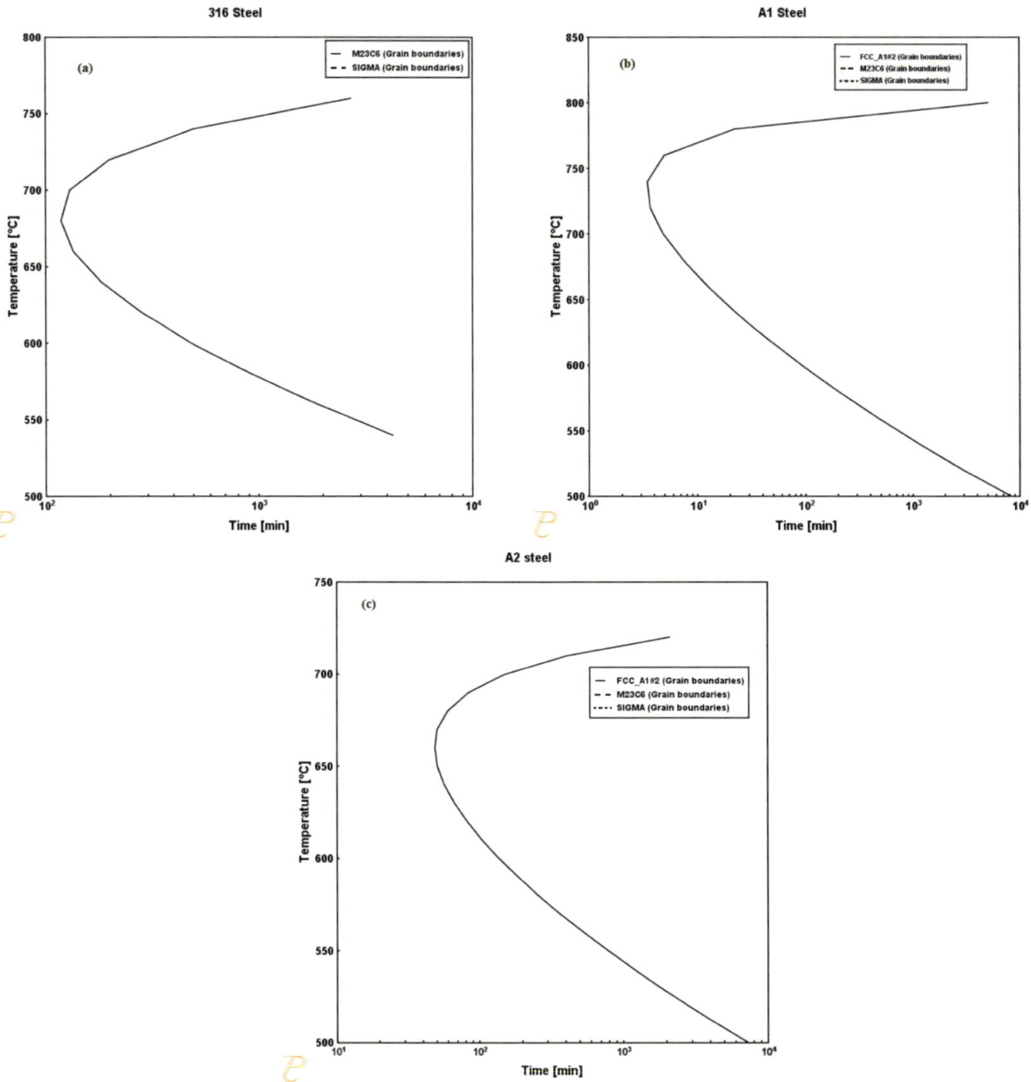

Figure 3. TTP diagram for (a) 316, (b) A1, and (c) A2 steels.

316 steel than that of A1 and A2 steels, which can be attributed to the lower content of interstitial solute causing a faster atomic diffusion process [2, 9]. The same reason can be adopted to explain the faster growth kinetics for aging of A1 steel. No σ phase and $M_{23}C_6$ formation is observed for these aging conditions. This behavior is in agreement with results reported in the literature [10, 11]. These diagrams show that the aging time for the intergranular carbide or nitride precipitation is very short; thus, this may be expected during the cooling at temperatures between 600 and 900°C in the welding process of the materials.

3. Microstructure evolution during aging

The SEM microstructures of the as-received 316, A1, and A2 steels are shown in **Figure 4(a–c)**, respectively. An austenite matrix is clearly observed for all the cases. Besides, neither intra-granular nor intergranular precipitation is observed for this condition.

Figure 4. SEM micrographs of as-received (a) 316, (b) A1, and (c) A2 steels.

Figure 5(a–f) illustrates, for instance, the SEM micrograph for 316, A1, and A2 steels after aging at 700°C for 10 and 1000 min, respectively. An intergranular precipitation can be observed on the austenite grain boundaries for all cases. The presence of intergranular precipitation is in good agreement with all the TTP diagrams. The highest and lowest volume fractions of intergranular precipitation correspond to the A2 and 316 steels, respectively. This is attributable to the highest content of interstitial solutes for the former steel. Conversely, the 316 steel has the lowest content of interstitial solutes. The volume fraction increases, in general with aging time. Nevertheless, the calculated growth kinetics is not in good agreement with the experimental precipitation observed in both the aged A1 and A2 steels, since the intergranular precipitation is present for aging times as short as 10 min. This fact can be attributed to the values of interfacial energy utilized in the TC-PRISMA calculation. In the case of 316 steel, the precipitation presence in observed at temperatures as high at 900°C for aging times up to 1000 min; however, the corresponding TC-PRISMA-calculated TTP diagram that shows a very slow kinetics of precipitation at 900°C.

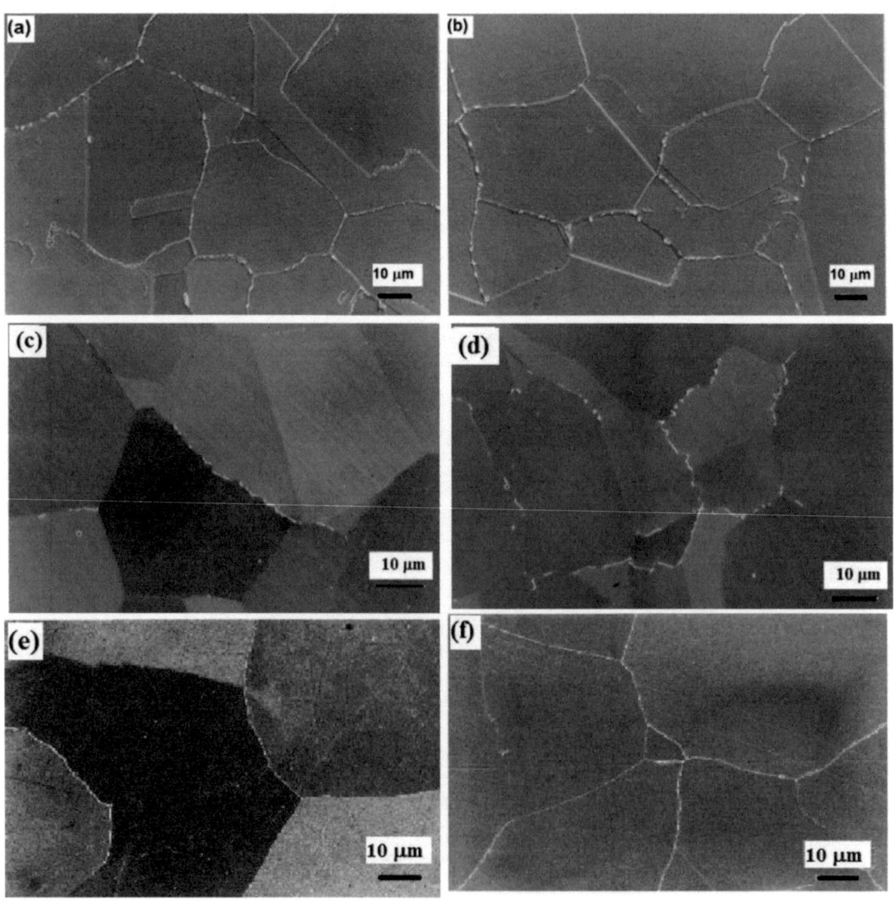

Figure 5. SEM micrographs of (a, b) 316, (c, d) A1, and (e, f) A2 steels aged at 700°C for 10 and 1000 min, respectively.

The XRD patterns of precipitates, extracted from electrolytic dissolution of the austenite matrix, are shown in **Figure 6(a, b)** for the 316 and A1 steels, respectively, aged at 700°C for 1000 min. These indicate that the precipitated phases correspond to $Cr_{23}C_6$, and $Cr_{23}C_6$ and Cr_2N for the aged 316 and A1 steels, respectively. The presence of $Cr_{23}C_6$ carbide is in good agreement with the TC-PRISMA-calculated TTP diagram for 316 steel; however, the N-containing steel shows the existence of both $Cr_{23}C_6$ and Cr_2N phases, which is not observed in the TC-PRISMA-calculated TTP diagram for the A1 steel. This can be also attributed to the interfacial energy between austenite matrix and carbide used for the TTP calculation. The nitrides and carbides detected in the XRD patterns are not only composed of Cr,

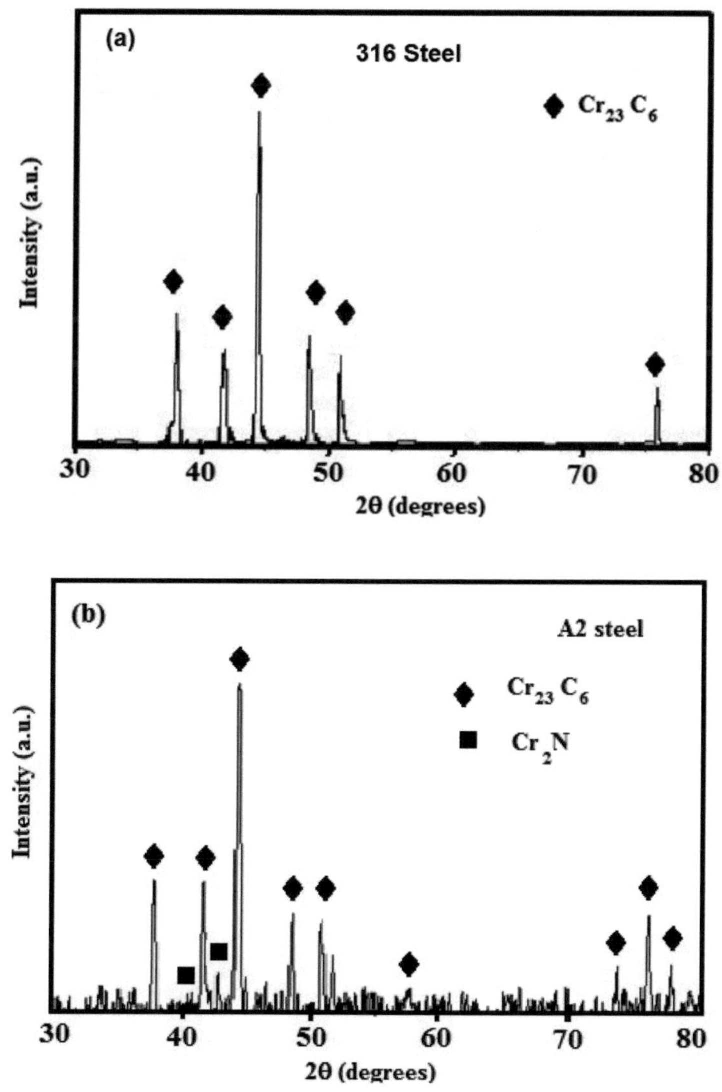

Figure 6. XRD pattern of (a) 316 steel and (b) A1 steel aged at 700°C for 1000 min.

but these are also formed by Fe and Mo according to the EDX-SEM analysis and compositions calculated by Thermo-Calc. Thus, they are designated as $M_{23}C_6$ and M_2N.

The experimental TTP diagrams, determined by SEM observations, for the three steels are shown in **Figure 7(a–c)**. The TTP corresponding to the 316 steel shows the intergranular precipitation of $M_{23}C_6$ carbides for all aging temperatures and times, **Figure 7(a)**, which is in good

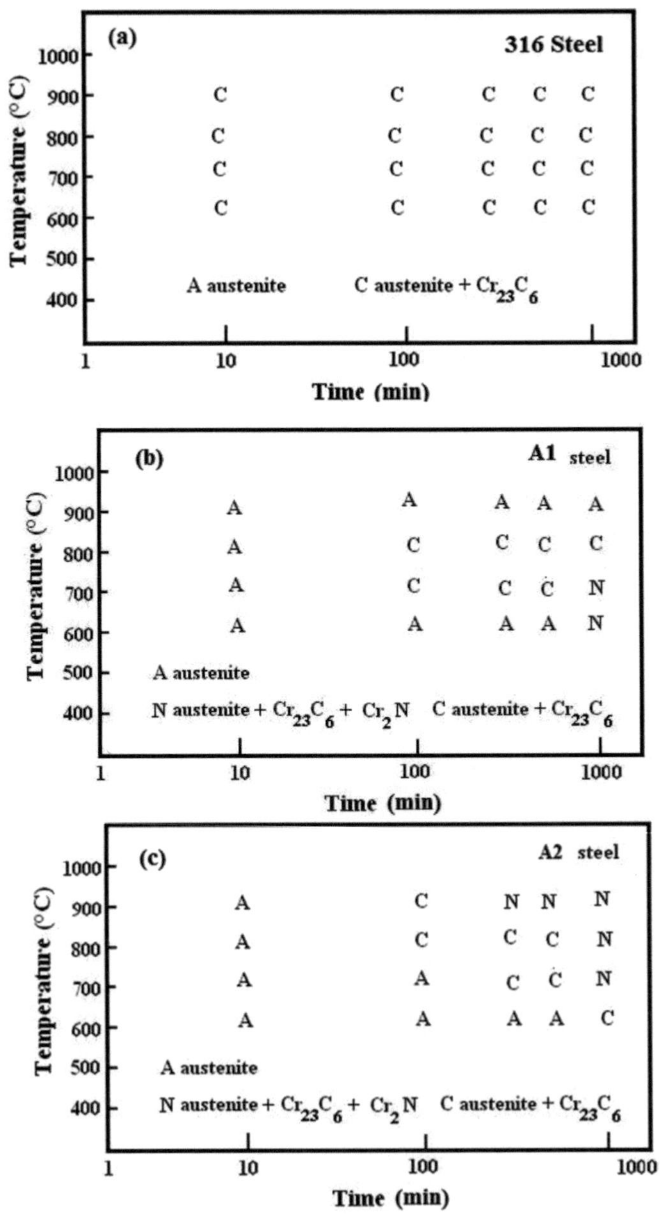

Figure 7. Experimental TTP diagram for (a) 316, (b) A1, and (c) A2 steels.

agreement with the calculated TTP diagram of 316 steel. That is, the growth kinetics of carbide is fast for all temperatures. The growth kinetics of carbide precipitation increases with aging temperature.

Figure 7(b) shows the experimental TTP for the A1 steel. This indicates that there is almost no precipitation at 600 and 900°C. The calculated TTP diagram for this steel, **Figure 3(b)**, exhibits no precipitation at 900°C, but there is nitride precipitation at aging temperature of 600°C. As mentioned above, this disagreement can be attributed to the values of interfacial energy between matrix and precipitate used for the TC-PRISMA calculation. In contrast, the experimental TTP diagram of the A2 steel, **Figure 7(c)**, presents precipitation at all temperatures being the slowest kinetics at 600°C, which agrees with the calculated TTP diagram, **Figure 3(c)**. It is important to note that the aging times for intergranular precipitation are very short at temperatures of 600–900°C; thus, the intergranular precipitation may be expected to occur during the welding process of thick structural sections. This fact may involve problems of sensitization or thermal aging.

4. Effect of precipitation on cryogenic toughness

Plots of CVN impact test energy at −196°C as a function of aging time for the three steel aged at different temperatures are shown in **Figure 8(a–c)**. In general, the impact energy decreases with aging time as a result of the increase in volume fraction of intergranular precipitates. In the case of the aged 316 steel, there is almost no decrease in impact energy for the specimens aged at 600 and 700°C, which is consistent with the low volume fraction of carbides detected at these temperatures. The highest decrease takes place for the specimen aged at 900°C, which observed the occurrence of the fastest growth kinetics of precipitation; and thus, the highest volume fraction of precipitates.

In the case of N-containing steels, the highest decrease in impact energy occurs at aging temperatures of 800 and 900°C for the A1 and A2 steels, respectively. The fastest growth of kinetics of precipitation was also observed to take place at those temperatures. No decrease in energy was observed practically to occur for aging of A1 steel at 900°C. In contrast, the decrease in energy was more drastic for aging of A2 steel as the aging time and temperature increase. This is attributable to the highest volume fraction of intergranular nitrides observed at this aging condition.

The SEM fractographs of CVN impact test specimens are shown in **Figure 9(a–f)** for the 316, A1, and A2 steel aged at 700°C for 0 (as-received) and 1000 min, respectively. The fracture mode was transgranular ductile, **Figure 9(a, b)**, for both the as-received and aged 316 steel specimens. This behavior suggests that the intergranular precipitation has a small effect on the toughness at cryogenic temperature. In contrast, the fracture mode of the aged A1 and A2 steels changes from transgranular ductile to intergranular brittle as the aging time increases, **Figure 9(c–f)**. The presence of intergranular brittle fracture can be more clearly noted in the aged A2 steels. This change in fracture mode can be attributed to the intergranular precipitation of nitrides.

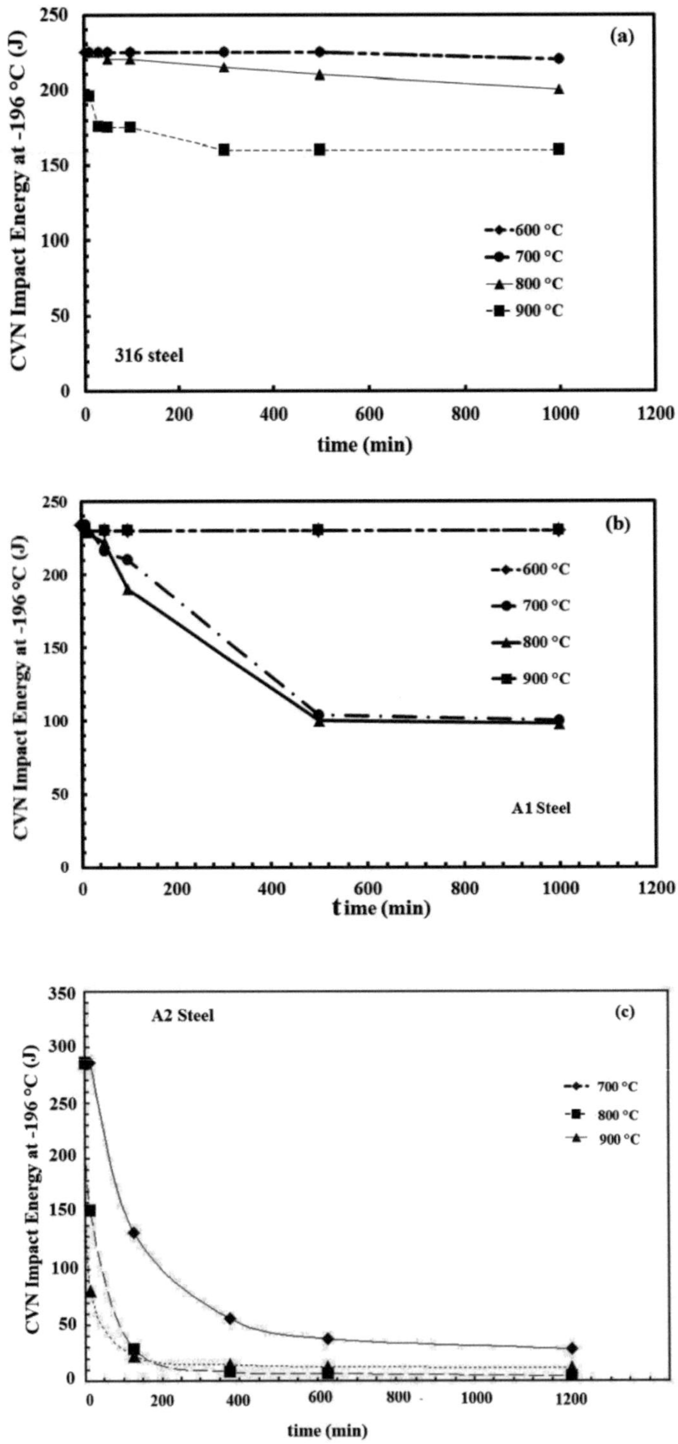

Figure 8. CVN impact energy at −196°C for the aged (a) 316, (b) A1, and (c) A2 steels.

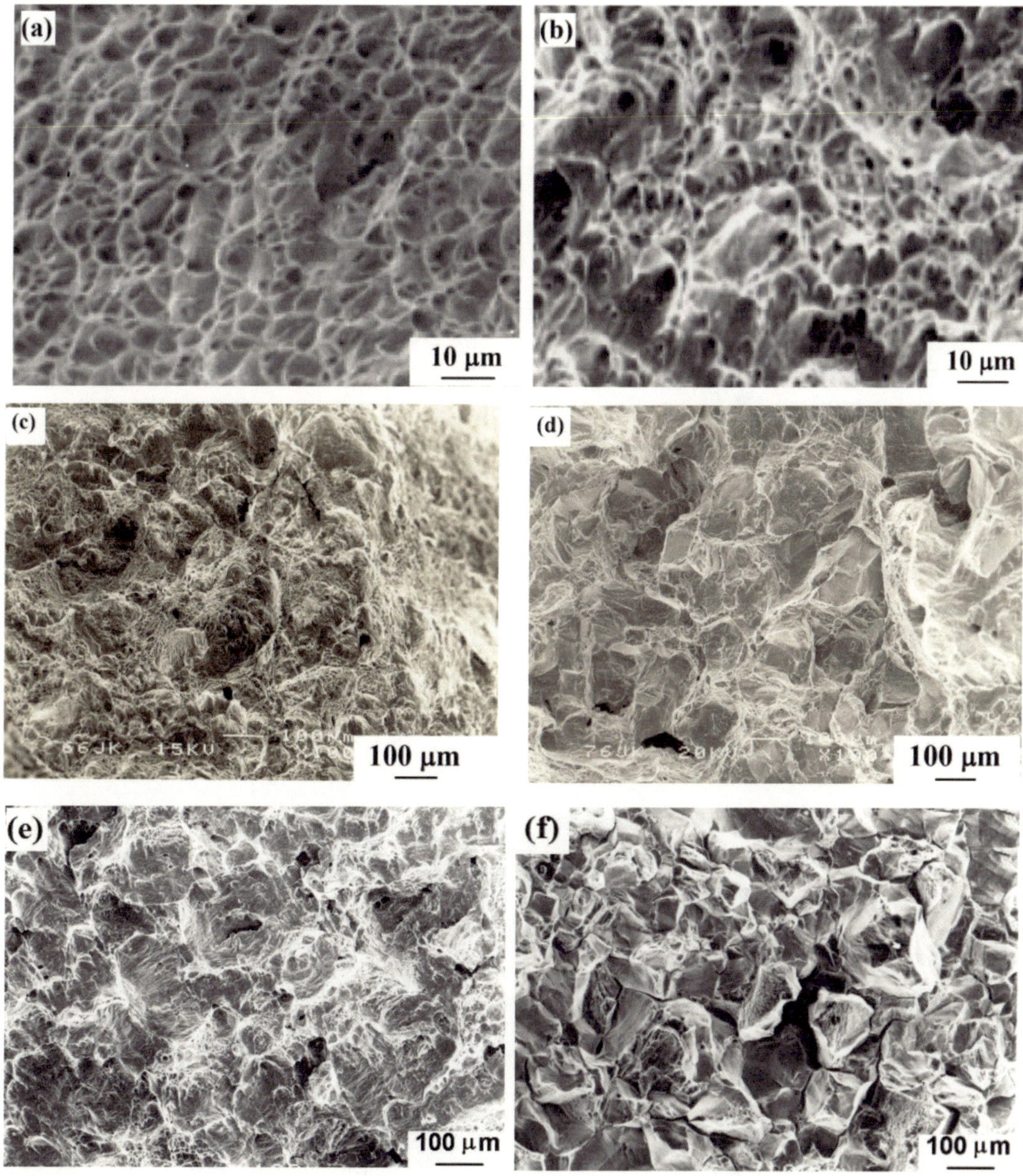

Figure 9. SEM fractographs of CVN impact specimens at −196°C for the (a, b) 316, (c, d) A1, and (e, f) A2 steels aged at 700°C for 0 and 1000 min, respectively.

The variation of CVN impact test energy at −196°C with the volume fraction of intergranular precipitates is shown in **Figure 10** for the aged A1 and A2 steels. These figures show clearly that the impact energy decreases with the increase in volume fraction of intergranular precipitates promoting the intergranular brittle fracture. This fact suggests that the intergranular precipitates cause the loss of cohesion of grain boundaries, as shown in **Figure 11**. This figure shows precipitate chains on grain boundaries for the A2 steel aged at 900°C for 10 min.

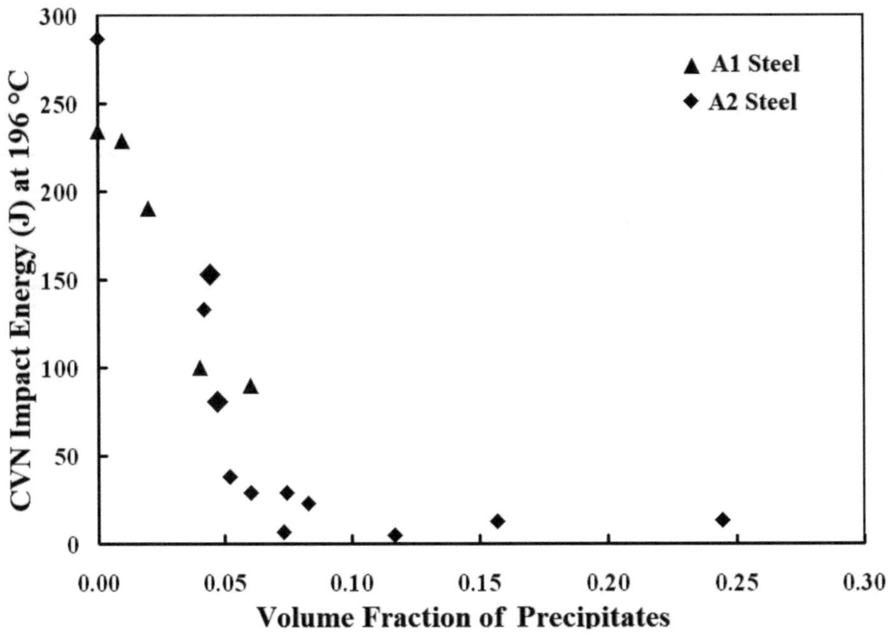

Figure 10. Plot of CVN impact test energy vs. volume fraction of precipitates for N-containing steels.

N-containing steels

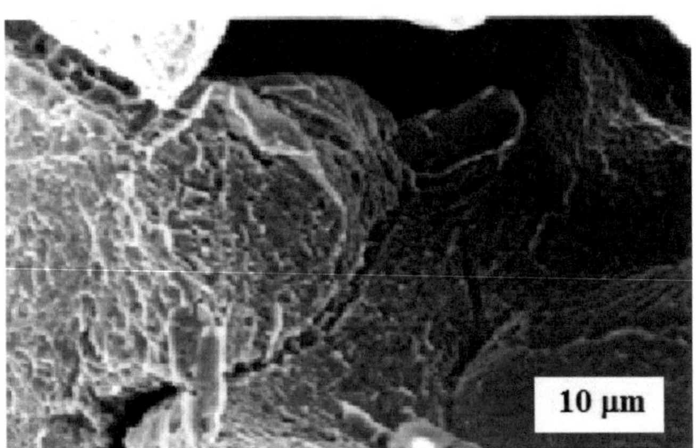

Figure 11. SEM fractograph of CVN impact test specimen of A2 steel aged at 900°C for 1000 min.

It is important to point out that the N-containing A1 and A2 steels have higher mechanical properties at cryogenic temperatures than those of the 316 steel; however, the former steels are more susceptible to the grain boundary embrittlement.

5. Summary

This chapter shows the effect of precipitation on cryogenic toughness for N-containing steels. The presence of intergranular nitride precipitates causes a severe decrease in CVN impact test energy for steels after aging at temperatures between 700 and 900°C for times as short as 10 min. This fact is attributable to its higher volume fraction of intergranular nitride precipitates due to their higher interstitial solute contents in comparison to 316-type steels. In contrast, the 316 steel shows a better resistant to the decrease in impact energy after aging in spite of the faster growth kinetics of precipitation. N-containing steels exhibit the presence of intergranular brittle fracture as a result of the grain boundary nitride precipitation. This behavior may be also present during the welding of this type of nitrogen-containing steels because of the short aging time for precipitation.

Acknowledgements

The authors wish to thank the financial support from Conacyt 220984 and 220929 and Instituto Politecnico Nacional-SIP.

Author details

Maribel L. Saucedo-Muñoz* and Victor M. Lopez-Hirata

*Address all correspondence to: maribelsaucedo@prodigy.net.mx

Instituto Politecnico Nacional (ESIQIE), CDMX, Mexico

References

[1] Marshall P. Austenitic Stainless Steels. 1st ed. London: Elsevier; 1984. 50 p

[2] Gavriluj VG, Berns H. High Nitrogen Steels. Berlin: Springer; 1999. 20 p

[3] Simmons JW, Atteridge DG, Rawers JC. Sensitization of high-nitrogen austenitic stainless steels. Corrosion. 1994;**50**:491-501. DOI: https://doi.org/10.5006/1.3294349

[4] Sedrinks AJ. Corrosion of Stainless Steels. 1st ed. New York: John Wiley and Sons Inc.; 1996. 13 p

[5] Stoter LP. Thermal ageing effects in AISI type 316 austenitic stainless steel. Journal of Materials Science. 1981;**16**:1039-1051. DOI: 10.1007/BF00542750

[6] Thermo-Calc. Thermo-Calc Software [Internet]. 2017. Available from:http://www.thermocalc.com/pro ducts-services/software/thermo-calc/ [Accessed: 2017-07-11]

[7] Thermo-Calc. Precipitation Module TC-PRISMA [internet]. 2017. Available from:http://www.thermocalc.com/products-services/software/precipitationmodule-(tc-prisma)/ [Accessed: 2017-07-11]

[8] Pettersson N., Frisk K., Fluch R. Experimental and computational study of nitride precipitation in a CrMnN austenitic stainless steel. Mat. Sci. Eng. A. 2017;**684**:435-441. DOI: http://dx.doi.org/10.1016/j.msea.2016.12.071

[9] Kostorz G. Phase Transformations in Materials. New York: Springer; 2001. 309 p.

[10] Saucedo-Muñoz ML, Watanabe Y, Shoji T, Takahashi H. Effect of microstructure evolution on fracture toughness in isothermally aged austenitic stainless steels for cryogenic application. Cryogenics. 2000;**40**:693-700. DOI: https://doi.org/10.1016/S0011-2275(01)00004-2

[11] Saucedo-Muñoz ML, Liu CS, Komazaki S, Kwon I, Hashida T, Takahashi H, Nakajima H. Evaluation of thermal aging embrittlement of austenitic stainless steels JN1, JJ1 and JK2 by cryogenic small-punch testing. Journal of Materials Research. 2002;**17**:852-858. DOI: http://dx.doi.org/10.1557/JMR.2002.0124

Chapter 7

Local Corrosion of Austenitic Steels and Alloys

Pisarevskiy Lev Alexandrovich and
Filippov Georgiy Anatolievich

Additional information is available at the end of the chapter

http://dx.doi.org/10.5772/intechopen.70339

Abstract

The principles of doping of austenitic unstabilized Cr-Ni and Cr-Mn-Ni steels resistant against pitting, slit, and intergranular corrosion (IGC) in weakly and strongly oxidative media are considered. The features of the effect of individual and joint doping with nitrogen, molybdenum, and silicon steels on their resistance to local corrosion have been studied. The ambiguous effect of joint doping with nitrogen and molybdenum of Cr-Mn-Ni steels on their resistance against pitting corrosion is revealed. It has been established that the presence of silicon, in addition to nitrogen and molybdenum, is a prerequisite for eliminating the propensity to the IGC after a long release of chromium-nickel steels. The introduction of silicon into Cr-Ni steel with 0.03% C, with a balanced content of Cr, N, Mo, and Si, equates it with a particularly low-carbon steel containing 0.003% C. Based on the above doping principles, a number of austenitic steels are designed for production equipment that is resistant to local corrosion in chloride-containing media and nitric acid.

Keywords: stainless steel, pitting corrosion, intergranular corrosion, crevice corrosion, weakly oxidative and strongly oxidative media environment, alloying, cold deformation, stable and unstable austenite, α'-phase, δ-ferrite

1. Introduction

The issues of reducing the metal capacity in industry, reducing the costs of repairing metal products, and increasing their service life and reliability are associated with the protection of the metal fund against corrosion. The use of stainless steels and alloys is an important direction in combating metal losses from corrosion in various industries and, above all, in those where metal products are used in corrosive aggressive environments. The greatest losses from corrosion are characteristic for the chemical industry and thermal and nuclear power.

In this case, the damage and failure of equipment in most cases are caused by the destruction of stainless steels as a result of their corrosion cracking, pitting, and intergranular corrosion.

In the chemical engineering industry, stainless chromium-nickel and chromium-nickel-molybdenum stabilized austenitic steels with a relatively high content of carbon 12Kh18N10T and 10Kh17N13M2T have been widely used as materials combining high strength with good processability in the metallurgical division. As a result of the development of industry and the increasing role of the economic factor, the shortcomings of these steels have also been revealed, in particular, low resistance against local corrosion in chloride-containing media. Therefore, low-carbon non-stabilized steels of 03Kh18N11, 03Kh17N14M3, and 03Kh19N10AГ3 types were developed and mastered by industry, which significantly increased the reliability of the welded equipment. At the same time, these steels are also prone to pitting corrosion with prolonged thermal exposures to the IGC, and their strength properties are even lower than those of 12Kh18N10T and 10Kh17N13M2T steels due to low-carbon content.

Increasing the resistance against local corrosion of austenitic steels is achieved as a result of reducing the content of carbon in them and doping with elements passivating the metal surface. It is known [1–3] that molybdenum inhibits active dissolution of steel, reduces the critical passivation current, and increases the critical temperature of pitting in chloride-containing media.

The use of nitrogen as an alloying element also makes it possible to obtain steels with high corrosion resistance. Like carbon, nitrogen forms a solid solution in the gland and promotes stabilization of the γ-phase. Like molybdenum, nitrogen reduces the propensity of steels to release intermetallic and carbide phases [3, 4].

The increased stability of the austenite of nitrogen-containing steels makes it possible to obtain from them highly corrosion-resistant low-magnetic products with a time resistance of the break of 2100–2150 N/mm^2 and a relative magnetic permeability of not more than 1.05 [5].

The introduction of nitrogen into the steel contributes to preventing or inhibiting the formation of border discharges in it and increasing its corrosion resistance [6]. However, the presence of only nitrogen or molybdenum in the metal is insufficient to completely eliminate the tendency to pitting corrosion [7, 8].

A high ability to self-passivate can have heat-treated steels alloyed together by nitrogen and molybdenum [2, 7–9]. Such steels in the cold-deformed state are also characterized by sufficient resistance to cracking during hydrogenation in chloride-containing media [10].

Increasing the resistance of austenitic steels against pitting corrosion is achieved by introducing silicon into them as an alloying element [11]. But silicon reduces the solubility of carbon, increases its thermodynamic activity, and accelerates the process of separating carbide phases, which can lead to a decrease in the resistance of the steel against intergranular corrosion [12].

In weakly oxidative environments, the influence of silicon on the propensity of austenitic steels to intergranular corrosion can be either positive or negative [13].

As follows from [14, 15], in steels Kh16N15M3 and Kh18N11 containing 0.03% C, an increase in the concentration of silicon leads to an increase in their propensity to the IGC in a weakly oxidative medium after tempering at temperatures above 650°C and to a decrease after heating below 650°C. Alloying with silicon in an amount of ≥3.29% of steel Kh20N20, containing not more than 0.032% C, suppresses its tendency to intergranular corrosion irrespective of the duration of tempering at 650°C within 1–100 h [16]. However, an increase in the silicon content to 5.40% in the same steel (with 0.015% C and 0.1% P) reduces its passivating ability [17].

The ambiguity of the influence of silicon on the corrosion resistance of tempered steels is caused by the formation of chromium-depleted border zones as a result of accelerating the release of excess phases, on the one hand, and facilitating the passivisability of these zones, thereby suppressing (complete or partially) their selective dissolution on the other. The resulting effect depends on which of these two factors is most prevalent [13].

Despite numerous studies, the effect of doping on the resistance to local corrosion of cold-deformed austenitic stainless steels with a stable and unstable austenite remains unclear.

2. Methodology of research

The investigated steels 22Kh18N5AM3D2S2 (EP 995) [18], 22Kh18N5AM3S2 (EP 996) [19], and 15Kh21G9N9AM2-Sh [20] were produced under industrial conditions. The first two were melted in 1-t induction furnace and the third in a 5-t electric arc furnace with a subsequent ESR. Forged bars with a cross section of 83 × 83 mm were rolled onto a wire rod with a diameter of 8 mm and then stretched onto billets with diameters of 3.0 and 1.5 mm. The workpiece with a diameter of 3.0 mm was quenched with heating from 1120 to 1150°C and cooling in water to study the microstructure and the resistance against IGC. Hardened billets with a diameter of 1.5 mm were pulled onto a wire 0.8 mm in diameter (deformation rate of 72%), from which samples were prepared for corrosion and mechanical and electrochemical tests.

The remaining meltings were produced in an induction furnace with a capacity of 40 kg. Forged blanks of 35 × 35 mm were rolled onto a wire rod with a diameter of 8 mm and stretched onto billets with diameters of 3.0 and 1.5 mm. After quenching from 1070 to 1150°C (depending on the content of carbon and nitrogen), 1.5-mm-diameter billets were drawn to a wire 0.8 mm in diameter, from which samples were prepared for the research.

Tests of steels for resistance against pitting and crevice corrosion were carried out under conditions of uniaxial tension equal to $0.8\sigma_b$. For this purpose cold-worked specimens were wound with wire of the same steel, passed into glass cylinders, filled with 10%$FeCl_3 \cdot 6H_2O$ solution (GOST 9.912-89, ASTM G48A), and loaded into adapted creep machines.

The anodic potentiodynamic polarization curves were taken at room temperature in a chloride solution of 1.0-n H_2SO_4 + 1.0-n NaCl by means of a potentiostat P-5848 in automatic mode at a polarization rate of 3.6 V/h using a standard silver chloride electrode, a computer IMM-IM3. Samples with a surface area of 5 cm^2 were mechanically polished, degreased, and activated with a current of 2×10^{-3} A/cm^2 for 5 min.

Resistance against IGC was tested on samples with a diameter of 3.0 mm after their tempering at 550–750°C for 1, 10, 100, and 500 h according to the standard method of AMU GOST 6032. Mechanical tests of cold-deformed wire samples were carried out in accordance with GOST 10446.

The microstructure was studied with a Neofot-2 light microscope. The composition of the phases was determined by the X-ray diffraction method; δ-ferrite and α'-phase (martensite of deformation) were detected from the value of the magnetic permeability of the samples, which was determined on a ballistic installation in a magnetic force field equal to 39.8×10^4 A/m (500 Oe).

3. Corrosion behavior of cold-deformed austenitic steels

Testing of samples under conditions of uniaxial tension in a 10% solution of $FeCl_3 \cdot 6H_2O$ made it possible to reveal the effect of the separate and combined alloying with nitrogen and molybdenum for unstabilized Cr-Ni and Cr-Mn-Ni cold-worked steels on pitting and crevice corrosion resistance. The results of the tests are shown in **Table 1**.

From the data obtained, it follows that steel without nitrogen and molybdenum, as well as separately doped with molybdenum or nitrogen, with both stable and unstable austenite, tends for pitting and crevice corrosion. These steels are destroyed after 7–30 h in contact with the wound wire because of the formation of deep pitting. The low resistance of samples against local corrosion is confirmed by the nature of the anodic potentiodynamic curves (**Figure 1**). The critical passivation current and the total passivation current, indicating the ability to pass into the passive state of 17Kh18N9 and 17Kh18AN10 steels, are maximum and are, respectively, $(2–5) \times 10^{-3}$ and $(1–5) \times 10^{-5}$ A/cm^2 (**Figure 1**, curves 1 and 2). On the polarization curves in the region of complete passivation, secondary activation maximum and unstable current jumps are seen associated with the formation of self-passivating pits.

Elimination of the tendency to pitting corrosion of cold-deformed Cr-Ni and Cr-Mn-Ni steels with a stable austenite structure (magnetic permittivity 1.05 Gs/Oe) is achieved, as can be seen from **Table 1**, when they are jointly doped with nitrogen and molybdenum. All nitrogen-doped chromium-nickel steels containing molybdenum do not collapse under prolonged stress tests. They are characterized by high resistance against pitting and crevice corrosion, which confirms the previously obtained data [8, 21, 22]. Areas of secondary activation absent on a polarization curve 3 steel 12Khl8N13AM3 (see **Figure 1**) are alloyed with nitrogen and molybdenum. The critical passivation current is minimal, and the region of potentials of total passivation is much wider than on curves 1 and 2 of steels 17Kh18N9 and 17Kh18AN10 (**Figure 1**). The surface of the 12X18H13AM3 steel samples remained light after the tests, without any trace of corrosion. The appearance of δ-ferrite and the formation of an α'-phase in the structure of nitrogen-containing Cr-Ni-Mo steels (1–3, see **Table 1**) do not reduce their resistance to local corrosion.

Austenitic steels with a super-equilibrium content of nitrogen after cold plastic deformation make it possible to obtain articles with yield strength of up to 3600 N/mm^2 or higher [23]. At

Steel (nominal melt number)	$\sigma_в$ (N/mm²)	Time to failure (h)	Phase composition	μ (Gs/Oe)	PRE
Cr-Ni steels					
22Kh18N5AM3D2S2 (1) (EP995)	2300	6500*	$\gamma + \alpha'$	7.5	30.3
22Kh18N5AM3S2 (2)	2670	6500*	$\gamma + \alpha'$	8.7	29.5
05Kh20N5AM2D2 (3)	1760	3000*	$\gamma + \delta + \alpha'$	15	29.2
03Kh18N9S2AM2 (4) 1920 3000* γ 1,05 28,0	1920 3000* γ 1,05 28,0	3000*	γ	1.05	28.0
03Kh17N13S2AM2 (5)	1915	1000*	γ	1.05	28.3
I2Khl8N13AM2 (6)	1815	1000*	γ	1.05	32.1
20Kh20N13 (7)	1610	20–30	γ	1.05	20.0
17Kh18AN10 (8)	1825	10–20	γ	1.05	21.8
07Khl8N10M3 (9)	1770	15–20	$\gamma + \alpha'$	10.1	27.4
17Khl8N9 (10)	1895	7–10	$\gamma + \alpha'$	7.5	17.8
Cr-Mn-Ni steels					
25Kh18G9N13M2 (11)	1820	10–15	γ	1.05	24.6
10Kh18G13AN4 (12)	1880	5–10	γ	1.05	22.1
25Kh18G9AN5M2 (13)	2240	1500*	γ	1.05	31.0
25Kh18G9AN5M2S2 (14)	2410	1500*	γ	1.05	31.2
15Kh21G9N9AM2-Sh (15)	2215	1500*	γ	1.05	35.8
10Kh18G9AN5M2S2 (16)	2000	5–10	$\gamma + \alpha'$	4.1	28.7
05Kh18G9AN5M2S2 (17)	1690	5–10	$\gamma + \delta + \alpha'$	7.3	29.1
05Kh18G9AN5M2S2 (18)	1650	1500*	γ	1.05	33.2

*Specimens taken from testing without failure.

Table 1. Properties, phase composition, and time to failure of cold-deformed wire specimens 0.8 mm in diameter of Cr-Ni and Cr-Mn-Ni steels in 10% $FeC1_3 \cdot 6H_2O$ solution with tensile stress of 0.8$\sigma_в$ under conditions, simulating pitting, and crevice corrosion.

the same time, it is known [24] that when they crystallize a large part of the nitrogen passes into nitrides of relatively large sizes. During subsequent heating, coarse particles of excess phases are retained within the structure, ineffective for steel strengthening and reducing its resistance to local corrosion.

In this regard, for the production of high-strength cold-deformed semifinished products, along with low-carbon steels with super-equilibrium nitrogen content, the use of nitrogen-containing steels with an increased carbon concentration, obtained by standard methods using nitrided ferroalloys, is effective. Studies are under way to test the compositions of corrosion-resistant Cr-Mn-Ni steels alloyed together by nitrogen and carbon [25], which significantly expands the γ-region on the equilibrium-phase diagram and improves the properties of steels [26].

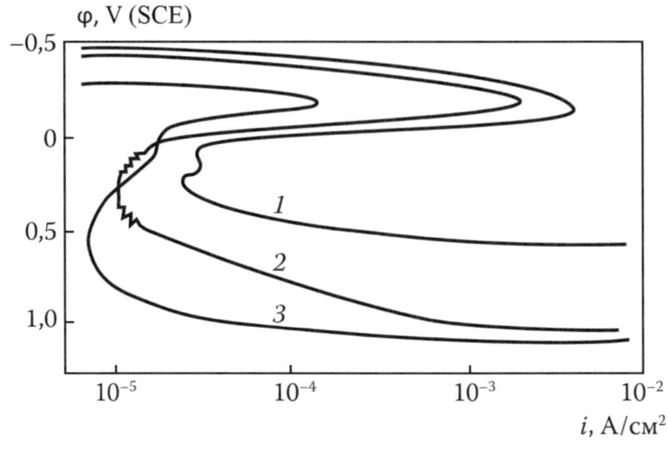

Figure 1. Anodic potentiodynamic polarization curves for cold-deformed steels 17Khl8N9 (1), 17Kh18AN10 (2), and 12Kh18N13AM3 (3) in 1.0-n H_2SO_4 + 1.0-n NaCl solution.

It was shown [27] that doping with copper facilitates the passivation of steel and increases its technological plasticity, ensuring the achievement of high strength due to large degrees of deformation [19, 28, 29].

The possibility of cold working of metals with high degrees of deformation is also ensured by sufficient solubility of excess phases in the structure of the steel during its high-temperature hardening (**Figure 2**), which simultaneously increases the corrosion resistance of the metal. On the polarization curve of 22Kh18N5AM3D2C2 steel (**Figure 3**, curve 1), there are no secondary activation regions; the current in the total passivation region does not exceed 10^{-5} A/cm^2, which characterizes a stable passive-state steel in a wide range of potentials. With a time resistance of 2300 N/mm^2, the steel 22X18H5AM3Д2C2 (EP 995) in a cold-deformed state is not prone to pitting and crevice corrosion and can withstand 6500 h without breaking in a chloride-containing medium at a stress of 0.8 $\sigma_в$ (**Table 1**).

In accordance with technical conditions developed, industrial batches of steel 22Khl8N5AM 3D2S2 (EP 995) have been prepared for corrosion-resistant marine cable structures 3 ×19(1 +9+9) and 3 ×7(1 +6) with an improved aggregate strength (16,200 and 13,100 N).

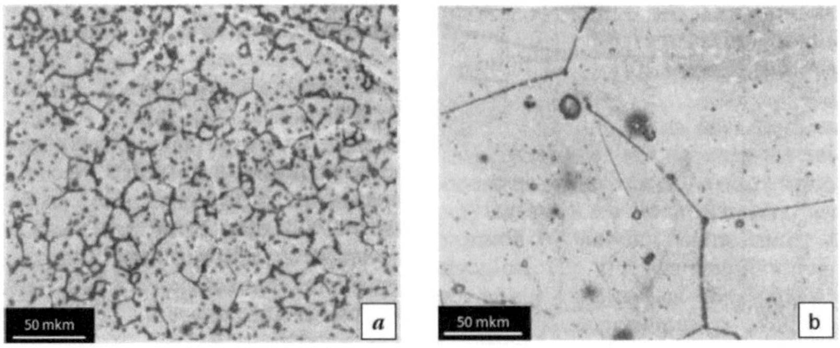

Figure 2. Microstructure of steel 22Kh18N5AM3D2S2 (EP 995) after water quenching from 1120°C (a) and 1150°C (b).

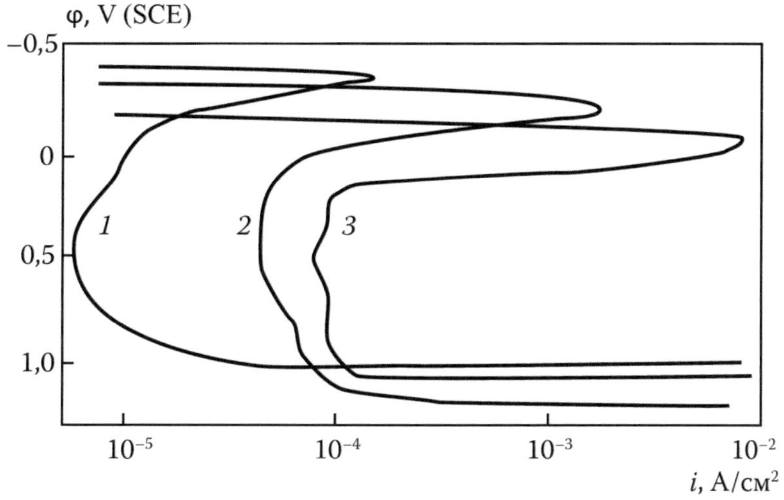

Figure 3. Anodic potentiodynamic polarization curves for cold-deformed steels 22Khl8N5AM3D2S2 (1) and15Kh21G9 N9AM2-Sh (2) in 1.0-n H_2SO_4 + 1.0-n NaCl solution and especially low-carbon alloy 02KhN40MB in chloride electrolyte after 100 h aging at 375°C (3).

As can be seen from **Table 1**, with separate doping with nitrogen and molybdenum, there is no significant increase in resistance to pitting corrosion of cold-deformed Cr-Mn-Ni steels. Samples of steels 25X18G9N13M2 and 10X18G13AN4 (melts 11 and 12) are destroyed after 5–15 h. Elimination of the propensity to local corrosion of Cr-Mn-Ni-like Cr-Ni steels is achieved by their joint doping with nitrogen and molybdenum. As a result, the samples survive the tests without failure for 1500 h (melting 13–15, 18). However, the high corrosion resistance of manganese steels, unlike chromium-nickel steels, is realized only if the former have a stable austenite structure, without the inclusion of δ-ferrite and α-phase, which increase the tendency to pitting corrosion and reduce the time to fracture of the samples to 5–10 h (melting 16 and 17).

As a whole, local corrosion resistance for cold-deformed manganese steels depends on the quantitative ratio within them of (δ+ α′)-phase manganese on the one hand and molybdenum, nitrogen, and chromium on the other. Therefore, due to the balanced content of the alloying elements, the steel 15X21Г9H9AM2-Ш [20] of industrial melting with the structure of stable austenite with the total content of nitrogen and carbon of 0.53% has a sufficiently high corrosion resistance. On a polarization curve (see **Figure 3**, curve 2), areas of secondary activation are absent. In the potential range from 0 to 900–1000 mV, the steel is in a stable passive condition, which points to the absence of a tendency toward pitting corrosion.

4. Potential resistance of austenitic steels to pitting corrosion

The potential resistance of steels to pitting corrosion can be estimated using the so-called PREN (pitting resistance equivalent number), calculated by the formula PRE = % Cr + 3.3 (%Mo) + 16 (%N) [30, 31]. It is believed that the higher the PRE, the more resistant the steel against pitting corrosion.

Table 1 shows the PRE values of the investigated steels, which, as a rule, correspond to their level of corrosion resistance. Elevated PRE values are characteristic for samples that do not break down during testing. Reduced values indicate the propensity of steels to pitting and crevice corrosion. At the same time, as can be seen from **Table 1**, melts 3 and 17, having the same structure and practically equal values of PRE, are different in resistance to local corrosion. One survives 3000 h of testing, and the other quickly collapses. In the corrosion-resistant Cr-Ni alloy 4, the PRE (28.0) is less than the Cr-Mn-Ni melting 17 (29.1), prone to pitting corrosion, which contradicts the meaning of the equivalent.

The obtained data indicate that the PRE formula does not take into account the different influence of the stability of austenite and inclusions of the second phases on the resistance to pitting corrosion of Cr-Mn-Ni and Cr-Ni steels doped with nitrogen and molybdenum, as well as their tendency to pitting corrosion in the case absence of one of these elements in the chemical composition of steel.

5. Pitting corrosion of nickel alloys

An analysis of the published data shows that nickel alloys not doped with N and Mo are prone to pitting corrosion (**Table 2**).

The Inconel 600 alloy containing a low percentage of chromium and high nickel has the lowest resistance to pitting corrosion and is prone, according to [32], to intergranular corrosion cracking.

Inconel 625 and Inconel 718 alloys, despite the high molybdenum content (up to 10 and 3.3%, respectively), under certain conditions, also have low resistance against pitting corrosion. Appearance of such corrosion of these alloys after aging in supercritical pressure (SCP) water at 500°C for 500 h is shown in **Figure 4** [33]. In this case, the PRE values reach the maximum values.

Alloy	C, ≤	Cr	Ni	Mo	N	PRE	Propensity to PC
Inconel 600	0.15	14–17	≤72	–	–	14–17	There is
Inconel 690	0.05	27–31	≤58	–	–	27–31	There is
Inconel 625	0.10	20–23	≤58	8–10	–	46–56	There is [33]
Inconel 718	0.08	17–21	50–55	2.8–3.3	–	26–31	There is [33]
02KhN40MB	0.02	18–20	30–40	4–5	–	31–36	There is [1]
KhN30MDB	0.03	29	31	3.5	0.05	41	There is [4]
					0.12	42	No [4]

Table 2. The content of basic alloying elements and nitrogen (wt.%) in nickel alloys and their propensity for pitting corrosion (PC).

http://dx.doi.org/10.5772/intechopen.70339

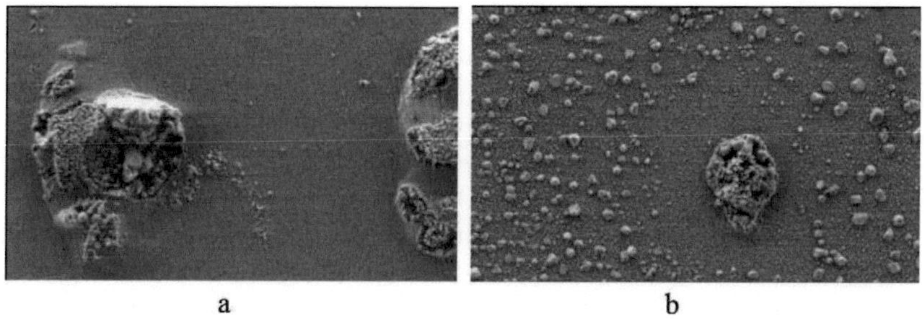

Figure 4. Pittings on the surface of nickel alloys Inconel 625 (a) and Inconel 718 (b) after their soak in water SCD at 500°C for 500 h.

The alloy ХН30МДБ (ЭК 77) [34], if its composition includes up to 0.05% N, also does not have sufficient resistance against pitting corrosion. However, an increase in the nitrogen content to 0.12% is sufficient to eliminate the propensity of the nickel alloy to pitting corrosion (**Table 2**) [4].

6. Intergranular corrosion of austenitic Cr-Ni steels

6.1. Increasing the resistance of steels against IGC in a weakly oxidative environment

The results of tests of samples in a solution of sulfuric acid and copper sulfate in the presence of copper chips according to the AMU GOST 6032 method after tempering at 650°C are given in **Table 3**. It shows that steel without Mo, doped with Si (1) or Si and N (2), is prone to IGC after tempering for 1–500 h [27]. The steels X16H15M3 [14] (3) and 03X20H19AM2 (4) alloyed together with nitrogen and molybdenum are corrosion-resistant after tempering only for 1 h. Alloying of these steels with silicon (see grade 03X17H13C2AM2 (5)) makes it corrosion-resistant after 100 H of heating.

Steel (nominal melt number)	C	Si	N	IGC resistance of steel tempered at 650°C (h)				
	Content (wt.%)			1	5	10	100	500
03Khl6NI4S2 [27] (1)	0.031	2.07	–	•	•	•	•	•
03Kh18N12AS2 (2)	0.029	1.53	0.16	•	•	•	•	•
Khl6N15M3 [14] (3)	0.032	≤0,4	0.12	○	•	•	•	•
03Kh20N19AM2 (4)	0.030	0.18	0.23	○	•	•	•	•
03Kh17N13S2AM2 (5)	0.034	1.71	0.20	○	○	○	○	•
03Kh17N13S2AM2 (6)	0.029	1.90	0.14	○	○	○	○	○
Khl6N15M3 [14] (7)	0.003	≤0.4	0.13	○	○	○	○	○

Light symbols, there is no IGC; dark symbols, there is IGC.

Table 3. Results of testing for IGC (AMU GOST 6032 procedure) resistance after tempering at 650°C steels with separate and combined alloying with N, Mo, and Si, containing ~0.03% C and steel Khl6N15M3 with 0.003% C.

Reducing the nitrogen content from 0.20 to 0.14% in this steel (6) eliminates its propensity to the IGC after a 500-h tempering. Similar corrosion resistance is possessed only by extremely low-carbon steel X16H15M3 (7), containing 0,003% of carbon [14].

Steel 03X17N13S2AM2 does not show a tendency to intergranular corrosion after tempering and other temperature and time parameters in the range 550–750°C (**Figure 5a**). This steel is much more resistant to IGC steel than the steel X16N15M3 with the same content of carbon and nitrogen (**Figure 5b**) [14], practically not inferior to especially low-carbon steel with 0.003% C (**Figure 5c**) and is not inclined to pitting and crevice corrosion (see **Table 1**, melting 5).

Thus, the introduction of silicon into the nitrogen-containing chromium-nickel austenitic steel with 0.03% C doped with molybdenum significantly improves its resistance against IGC and at a balanced content of Cr, N, Mo, and Si equates to a particularly low-carbon stainless steel containing 0.003% C.

The results of the investigations show that improvement of passivability and elimination of the propensity to pitting corrosion of cold-deformed austenitic chromium-nickel and stably austenitic chromium-manganese nickel steels with high- and low-carbon content are achieved by their joint doping with nitrogen and molybdenum.

However, to prevent a tendency for the IGC to leave after tempering in the investigated temperature-time range of stainless steels, their passivation with doping with nitrogen and molybdenum is insufficient. Apparently, this is due to the fact that when heated in the region of dangerous temperatures along the grain boundaries, chromium-depleted zones are formed as a result of the precipitation of excess phases. Therefore, the necessary condition for stability against IGC in the weakly oxidative environment of austenitic Cr-Ni steels after their long tempering at 550–750°C is the simultaneous presence in them of not only nitrogen and molybdenum but also a sufficient amount of silicon that improves, according to [13, 16, 17], the passivation of chromium-depleted boundary zones and reduction in tens or even hundreds of times the dissolution rate of tempered steels containing ≤ 0.032% C.

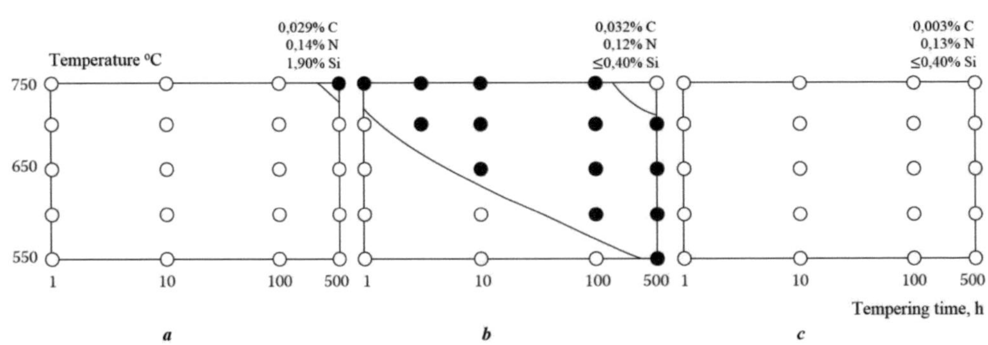

Figure 5. Resistance against IGC of nitrogen-containing steels 03Khl7N13S2AM2 (a) and Kh16N15M3 (b, c) with different contents of C and Si. Light symbols, there is no IGC; dark symbols, there is IGC.

6.2. New steels resistant to IGC in a strongly oxidizing environment

The main structural material for the manufacture of equipment producing nitric acid is steel 03Kh18N11 and its analog AISI 304L. Nitric acid is a strong oxidant, and the equipment used in its production has an inadequate service life due to corrosion damage.

In TsNIIchermet developed austenitic steels of the new system of doping Cr-Ni-N-Si (≤0.03% C, 14–17% Cr, 9–11% Ni, 2–4.5% Si) are intended for work in contact with highly oxidizing media [35–37]. The development is based on the nitrogen-containing steel 03Kh17AN9, which was doped with silicon in an amount that provides the formation of an austenite structure, and also the content of chromium and nickel was changed. The concentration of nitrogen and silicon in the steels was in the range of 0.08–0.145 and 2.0–4.5%, respectively.

The most positive results were obtained on the basis of two doping systems, (14–16)% Cr–11% Ni and (14–17)% Cr–9% Ni, with additions of nitrogen and silicon.

The metal for investigation was melted in a vacuum induction furnace and poured into 10 kg ingots that were forged on bobbins and rolled onto a 10-mm-thick sheet. The samples were quenched in water after heating to 1050°C. Resistance of steels against IGC was determined on polished cylindrical specimens with a diameter of 5 mm, a length of 60 mm according to a method simulating the method of DU GOST 6032. An analogue express method was used in which by using a more oxidizing medium, 27% HNO_3 + 40 g/l Cr^{6+} reduces the test time to 10 h.

The results of tests of steels for resistance against IGC are presented in **Table 4**. It can be seen that in comparison with steel 03Kh18N11, a decrease in chromium content from 16 to 14% at a nickel concentration of 11% and joint doping with nitrogen and silicon contributes to a 5–10-fold decrease in the mass loss of samples, and a decrease in chromium content from 17 to 14% and nickel to 9% is 5–28 times.

Steel	**Test time (h)**			**$\Delta m_{st}/\Delta m$**
	1	**6**	**10**	
	Mass loss Δm (g)			
03Kh18N11	0.014	0.327	1.046	–
03Kh16N11AS2	0.014	0.095	0.212	4.93
03Kh15N11AS2	0.015	0.078	0.127	8.24
03Kh14N11AS2	0.015	0.062	0.102	10.25
03Kh17N9AS2	0.012	0.091	0.194	5.39
03Kh15N9AS2	0.016	0.099	0.196	5.34
03Kh14N9AS4	0.002	0.021	0.037	28.27

Note: Δm_{st} и Δm, mass loss of standard steel 03Kh18N11 and new steels for 10 h.

Table 4. The change in the mass of the samples as a function of the time of testing for IGC in a strongly oxidizing medium.

The increase in resistance against intergranular corrosion in a highly oxidative environment of nitrogen-containing silicon-doped Cr-Ni steels should be attributed to its ability to enrich the surface layers, thereby increasing the protective properties of the passivating films.

In accordance with the new system of doping of Cr-Ni-N-Si, new austenitic steels 03Kh17N9AS2 and 03Kh14N9AS4 are developed that are not inclined to IGC in a strongly oxidative environment and are much more corrosion-resistant than standard steel 03Kh18N11.

It is generally accepted that the main cause of the destruction of steels under conditions of contact with an aggressive medium is local corrosion of the metal, which leads to a decrease in its working section. However, there are reasons to believe that another reason for the cracking of the metal can be its hydrogen saturation due to the development of delayed fracture phenomenon [38–41]. The results of the study showed that the new steels 03Kh17N9AS2 and 03Kh14N9AS4, designed for operation in highly oxidative media, also have an increased resistance to delayed fracture and hydrogen embrittlement [42, 43].

7. Local corrosion and choice of candidate steel for heat-exchange tubes of NPP with WWER

Historically, the steel 08X18H10T was chosen for heat-exchange pipes of the first domestic steam generators of NPPs with WWER. Despite all subsequent upgrades, this steel remained unchanged for steam generators [44]. While during this time at foreign nuclear power plants

Steel/alloy	NPP with	$\sigma_{0.2}$	$\sigma_{в}$	δ (%)	PRE	Resistance against the PC (○); inclination to PC (•)	Resistance against the IGC (○); inclination to IGC (•)
		N/mm²					
		Not less than					
Applied steels and alloys							
AISI 304*	PWR	205	515	40	19	•	•
AISI 316*		205	515	40	25	•	•
Inconel 600*		205	550	35	15.5	•	•
Inconel 690		240	550	45	29	•	○ •
08Kh18N10T	WWER	196	490	40	18	•	○ •
Candidate steels							
03Kh17N13S2AM2	WWER	340	650	40	30	○	○
AISI 316LN		280	600	40	28	○	○•
AISI 316L [30]		170	485	40	26	•	○
AISI 316 Ti [30]		220	520	40	24	•	○

*Currently not applicable.

Table 5. Mechanical properties and resistance to pitting (PC) and intergranular (IGC) corrosion of steels and alloys used previously and currently for the production of heat-exchange tubes of nuclear power plants (NPP) with PWR and WWER.

with PWR, not a single generation of pipe system materials has been replaced [30]. The AISI 304 and 316 steels were replaced with nickel alloys—first, it was Inconel 600 and then the more corrosion-resistant Inconel 690 (see **Table 5**). However, recently this alloy also becomes a problem for NPPs with PWR. The instability of its corrosive behavior is close to the characteristics of the Inconel 600 alloy [45].

There is every reason to believe that the problems of damage to heat-exchange pipes made of AISI 304 and 316 steels and Inconel 600 and 690 alloys have led to ignoring their propensity to local corrosion (**Table 5**). In this respect, steel 08Kh18N10T is also rather mediocre. It is prone to pitting corrosion [30] and, in certain cases, to the IGC.

Steel X16N15M3 [14] (see **Table 2**, steel 3) is an analog of AISI 316LN and is not prone to pitting corrosion (**Table 5**), but is resistant to MCC after tempering at 650°C for only 1 h.

Thus, among the candidate materials for heat-exchange pipes, the most promising, as can be seen from **Table 5**, is steel 03Kh17N13S2AM2, the chemical composition of which is developed on the basis of grade 03Kh18N9S2AM2 [46]. Steel has the highest mechanical properties and no tendency to local corrosion.

8. Corrosion resistance of steel for heat-exchange tubes of reactor with lead coolant

The new steel 03Kh18N13S2AM2VFBR-Sh (EP 302M-Sh) [47], designed on the basis of grades 03Kh18N9S2AM2 [46] and 10Kh15N9S3B-III, was proposed for manufacturing heat-exchange tubes of the steam generator of the said reactor. Pipes, made of steel 03Kh18N13S2AM2VFBR-Sh, will be operated simultaneously in liquid lead and chloride-containing steam-water medium.

The results of her tests, presented in **Table 6**, indicate that the rate of crevice and pitting corrosion of this steel is 2 orders of magnitude lower than the standard grades 08Kh18N10T, stainless in an aqueous medium, and 10Kh15N9S3B-Sh (EP 302-Sh) designed for operation in liquid lead. In tests for stress-corrosion cracking, samples from a new steel—Odinga rings—do not collapse, in contrast to standard steels, for 1000 h (**Table 7**).

Steel	SC rate	PC rate
	g/m²h	
03Kh18N13S2AM2VFBR-Sh	0.10	≤0.01
10Kh15N9S3B	25	9.6
08Kh18N10T	16	7.8

*On contact with a sealing sulfur-containing material.

Table 6. The slit (SC)* and pitting (PC) corrosion rate in 10% $FeCl_3 \cdot 6H_2O$ solution of the new 03Kh18N13S2AM2VFBR-Sh and standard steels.

03Kh18N13S2AM2VFBR-Sh	10Kh15N9S3B	08Kh18N10T
Tensile stress (N/mm^2)		
523	552	295
Time to failure (h)		
1000*	43	10
*Samples are not destroyed.		

Table 7. Time before the destruction of the Odinga rings in 10% $FeCl_3 \cdot 6H_2O$ solution with tensile stress of $1.2\sigma_{0.2}$ of the new 03Kh18N13S2AM2VFBR-Sh and standard steels.

Corrosion indicators	03Kh18N13S2AM2VFBR-Sh	12Kh18N10T	AISI 316L
Weight gain (g/m^2)	0.33	0.54	–
Corrosion rate (mg/m^{2}h)	0.88	1.5	5.6

Table 8. Corrosion parameters of polished samples of steel 03Kh18N13S2AM2VFBR-Sh, 12Kh18N10T in an aqueous medium at 505°C, pressure 17 MPa for 1000 h [39], and AISI 316L steel at 480°C, 25 MPa pressure for 500 h [49].

Table 8 shows the results of gravimetric measurements of polished samples of steel 03Kh18N13S2 AM2VFBR-Sh after 1000 h of testing in an aqueous medium of an autoclave complex at 505°C, a pressure of 17 MPa, and a dissolved oxygen concentration of 30 ppb [48]. From the data obtained, it follows that the corrosion rate of the new steel in the water-coolant environment is almost 1.5 times lower than that of the grade 12Kh18N10T and several times smaller than that of the AISI 316L steel. The new steel after the tests remained without trace of corrosion, unlike the Inconel 625 and 718 alloys with pitting on the surface of the samples after they had been held in supercritical pressure (SCP) water with an oxygen concentration of 25 ppb at 500°C for 500 h (see **Figure 4**).

9. Conclusion

Results of investigation of the effect of joint and separate doping with nitrogen and molybdenum, cold deformation, inclusions of the α'-phase, δ-ferrite on the tendency to pitting corrosion of Cr-Ni and Cr-Mn-Ni steels with stable and unstable austenite, studying the influence of silicon, and nitrogen resistance to intergranular corrosion of Cr-Ni steels are essentially the principles of alloying unstabilized austenitic steels that are resistant to local corrosion in weakly and strongly oxidizing environments. On their basis, a number of austenitic steels of the following types have been developed: 22Kh18N5AM3D2S2 (EP 995), 22Kh18N5AM3S2 (EP996), 15Kh21G9N9AM2-Sh, 03Kh18N9S2AM2, 03Kh17N13S2AM2, 03Kh18N13S2AM2VFBR-Sh, 03Kh17N9AS2, 03Kh14N9AS4, 03Kh20N19AM2 (EC 176), 12Kh18N13AM3 (EP 878), etc. These steels are designed to produce equipment that is resistant to local corrosion in chloride-containing media and nitric acid.

Author details

Pisarevskiy Lev Alexandrovich and Filippov Georgiy Anatolievich*

*Address all correspondence to: iqs@bk.ru

Bardin Central Research Institute of Ferrous Metallurgy (TsNIIchermet), Moscow, Russia

References

[1] Ulyanin EA, Svistunova TV, Levin FL. Highly Alloyed Corrosion-Resistant Alloys. Moscow: Metallurgiya. 1987:39-41

[2] Feldgandler EG, Sorokina NA, Svistunova TV, Levin FL. Modern corrosion-resistant steels and alloys for media of different aggressiveness. In: Problems of Modern Metallurgy. Moscow: Metallurgiya; 1983. pp. 160-168

[3] Sokol Ya I, Ulyanin EA, Feldgandler EG. Structure and Corrosion of Metals and Alloys: Atlas. Directory. Moscow: Metallurgiya; 1989. pp. 77

[4] Shlyamnev A.P. Nitrogen-containing stainless steels—Structure, properties, production and application prospects. Problems of Ferrous Metallurgy and Materials Science. 2007;**1**:53-60

[5] Pisarevsky LA. Malomagnetic steels for high-strength wire. In: Structure and Properties of Non-Magnetic Steels. Moscow: Science; 1982. pp. 194-197

[6] Gulyaev AP, Pisarevsky LA. Investigation of properties and industrial testing of a new high-strength low-magnetic steel. In: High-Strength Non-Magnetic Steels. Moscow: Science; 1978. pp. 186-191

[7] Chigirinskaya LA, Chernova GP, Tomashov ND. Corrosion behavior of chromium-nickel-molybdenum steel with nitrogen. Protection of Metals. 1984;**XX**(3):404-407

[8] Pisarevsky LA, Golovanenko SA. Structure and properties of stainless steels with nitrogen and molybdenum. Sat. Thesis dokl. Conf. "New in Metallurgy and Heat Treatment of Metals and Alloys". Chelyabinsk: NIIM; June, 1983. pp. 6-8

[9] Golovanenko SA, Pisarevsky LA. Corrosion-resistant steels for high-strength cables and wires. In: Problems of Modern Metallurgy. Moscow: Metallurgia; 1983. pp. 153-159

[10] Pisarevsky LA, Aparin DV, Ulyanin EA, et al. Effect of nitrogen and molybdenum on the resistance of non-magnetic steels against hydrogen cracking. In: New Technologies for the Production of Special Steels. Moscow: TsNIIChM, Electrostal Plant; 1990. pp. 133-136

[11] Glazkova SA, Shvarts GL, Freiman LI, Tavadze FN. Investigation of the stability of chromium-nickel steels alloyed with molybdenum and silicon, to local corrosion. Protection of Metals. 1974;**10**(1):9-16

[12] Svistunova TV, Sakuta ND. Effect of carbon, silicon and phosphorus on the mechanical properties of the alloy NM27. In: Improving the Characteristics of High-Quality Steels due to Optimization of Alloying and Structure. Moscow: Metallurgia; 1984. pp. 33-37

[13] Kasparova OV. Features of intercrystalline corrosion of siliceous austenitic stainless steels. Protection of Metals. 2004;**40**(5):475-481

[14] Savkina Ya L, Feldgandler EG. Effect of alloying on the tendency to intercrystalline corrosion of steel 000X16H15M3. MITOM. 1968;**11**:10-13

[15] Lozovatskaya LP, Levin IA, Burtseva IK, et al. Increase in the resistance of 03X18H11 steel against the MCC by adjusting its chemical composition. Protection of Metals. 1984;**20**(3):411-415

[16] Kasparova OV, Milman VM, Kostromina SV. On the mechanism of the effect of silicon on the intergranular corrosion of tempered austenitic stainless steels. Protection of Metals. 1991;**27**(1):55-63

[17] Kasparova OV, Baldokhin Yu V. Effect of silicon on the electronic structure and corrosion-electrochemical behavior of phosphorus-containing steel Kh20N20. Protection of Metals. 2002;**38**(5):463-469

[18] Pisarevsky LA, Golovanenko SA, Rudchenko AV, et al. Corrosion-Resistant Steel. 10.02.1981, publ. 01.07.83; A. with. 986957, IPC^3 C22C38/54

[19] Golovanenko SA, Pisarevsky LA, Melkumov IN, et al. Corrosion-Resistant Steel. 10.12.1979, publ. 30.05.81; A. with. 834223, IPC3 C22C38/44

[20] Pisarevsky LA, Aparin DV, Ulyanin EA, et al. Corrosion-Resistant Steel. 10.09.1989, publ. 30.11.91; A. with. 1694685, IPC5 C22C39/58

[21] Pisarevsky LA, Filippov GA, Lipatov AA. Effect of N, Mo and Si on local corrosion resistance of unstabilized Cr-Ni and Cr-Mn-Ni-austenitic steels. Metallurgy. 2016;**8**:59-66

[22] Pisarevsky LA, Kaputkin II. Influence of thermal treatment of nitrogen-containing corrosion-resistant steel on its structure and local corrosion. In: Optimization of Alloying and Heat Treatment of High-Quality Steels. Moscow: Metallurgiya; 1987. pp. 31-33

[23] Berezovskaya VV, Valiev RZ, Sokolovskaya Yu A. Mechanism of plasticity and phase stability of nitrogen-doped austenite in Cr-Mn-Mo steel under intense plastic deformation. Vestnik TSU. 2013;**18**(4):1961-1962

[24] Kaputkina LM, Svyazhin AG, Prokoshkina VG. Dissolution and separation of excess phases and nitrogen distribution between a solid solution and nitrides in corrosion-resistant steel. Metals. 2006;**5**:93

[25] Berezovskaya VV. The system of doping of high-nitrogen austenitic steels, structure, mechanical and corrosive properties. Innovations in Materials Science and Metallurgy: Materials of I International. Interactive. Scientific-practical. Conf. [13-19 Dec. 2011, Yekaterinburg]. Ekaterinburg: Publishing house Ural. Un-ta; 2012. Part 1. pp. 257-266

[26] Berns H. High interstitial stainless austenitic steels, part I: Constitution, heat treatment, properties, applications. Proceedings of the 10th International Conference on High Nitrogen Steels. HNS 2009, Moscow, Russia: Moscow, MISIS; 2009. pp. 129-139

[27] Feldgandler EG, Plaskeev AV. Influence of doping of Si and Cu on corrosion-electrochemical and mechanical properties of austenitic steel. MITOM. 2003;**10**:12-21

[28] Pisarevsky LA, Kaputkin II, Bogatova NL, et al. The structure and properties of the new corrosion-resistant steel 30Х18Н4АМ3Д2С2: В кн. Influence of Alloying and Heat Treatment on the Properties of High-Quality Steels and Alloys. Moscow: Metallurgiya; 1985. pp. 30-33

[29] Kaputkina LM, Prokoshkina VG, Svyazhin AG, et al. The structure and properties of stainless steel doped with nitrogen and copper. MITOM. 2009;**6**(648): 23-28

[30] Karzov GP. Problems and prospects for the use of structural materials for horizontal steam generators. Teploenergetika. 2011;**3**:22-26

[31] Metals Handbook Desk Edition. 2nd ed. ASM International. The Materials International Society; 2006

[32] Baranenko VI, Oleinik SG, Belyakov OA, et al. Influence of the choice of structural materials on the damageability of heat-exchanging tubes of steam generators of nuclear power plants with PWR and WWER. Izvestiya Vuzov. Nuclear Energy. 2004;**2**:2-13

[33] Allen TR, Sridharan K, Chen Y, Tan L, Ren X, Kruizenga A. Research and Development on Materials Corrosion Issues in the Supercritical Water Environment: PREPRINT—ICPWS XV Berlin, September 8-11, 2008

[34] Pat. 2288967 RF, C22C38/54. Corrosion-resistant alloy and an article made of it/Svistunova TV, Shlyamnev AP, Litvak BS, Danovskiy NG Claimed. 15.04.2005; Publ. 10.12.2006

[35] Shlyamnev AP, Filippov GA, Naumenko VV. Properties of austenitic chromium-nickel steels doped with nitrogen and silicon. MITOM. 2011;**2**:22-26

[36] Naumenko VV, Shlyamnev AP, Filippov GA. Mechanical and corrosion properties of stainless steels of the Fe-Cr-Ni-N-Si alloying system. Problems of Ferrous Metallurgy and Materials Science. 2010;**4**:68-74

[37] Shlyamnev AP, Uglov VA, Filippov GA, et al. Stainless steels with nitrogen: Structure, properties, issues of production technology. Bulletin of Scientific, Technical and Economic Information. 2013;**1358**(2):42-57

[38] Filippov GA, Sarrak VI, Perkas MD. The phenomenon of delayed destruction of martensitizing steel. DAN USSR. 1976;**226**(4):83-86

[39] Filippov GA, Sarrak VI. The local distribution of hydrogen and internal microstresses in the structure of hardened steel. FMM. 1980;**49**:121-125

[40] Mishin VM, Filippov GA. Physics of delayed destruction of steels. Mineral Waters: Polygraphprom. 2013:455

[41] Mishin VM, Filippov GA. Criterion and physicomechanical characteristic of resistance become delayed destruction. Deformation and Destruction. 2007;**3**:37-42

[42] Naumenko VV, Filippov GA, Shlyamnev AP, Shabalov IP. Investigation of the durability of low-carbon nitrogen-containing corrosion-resistant steels to hydrogen embrittlement and delayed destruction. Steel. 2013;**6**:58-63

[43] Alekseeva LE, Naumenko VV, Filippov GA, Shabalov IP. The effect of nitrogen and silicon on the stability of austenite and the propensity to decelerate destruction of low-carbon austenitic stainless steels of the Fe-Cr-Ni-N-Si system. Problems of Ferrous Metallurgiya and Materials Science. 2012;**2**:55-60

[44] Karzov GP, Suvorov SA, Fedorova VA, et al. The main mechanisms of damage to heat-exchanging pipes. The Sixth International Workshop on Horizontal Steam Generators. Podolsk; March 2004

[45] Bergunker VD. Problems of material selection for the RM with PWR. Alloy 690 TT. Materials of the Conference OKB. Podolsk: Hydropress; May 17-20-2011

[46] Pisarevsky LA, Aparin DV, Ulyanin EA, et al. Corrosion-Resistant Steel. 08.05.1988, publ. 15.06.90; A. with. 1571099, IPC5 C22C39/44

[47] Pat. 2499075 RF, C22C38/58 C22C38/54. Corrosion-resistant austenitic steel/Bozin SN, Lemehov VV, Novichkova OV, Pisarevsky LA, Filippov GA. Claimed. 21.08.2012; Publ. 11.20.2013

[48] Artjukhova TN, Aleshin AA, Belous VN, et al. Investigation of the corrosive behavior of steel grade EP 302M-SH in a medium of a water coolant under operating conditions of a steam generator RU BREST-OD-300//Sb. Theses of the Reports of the Thirteenth International Conference "Problems of Materials Science in the Design, Manufacture and Operation of Nuclear Power Plant Equipment". St. Petersburg, Russia: CRI KM Prometey; June 2014. pp. 100-101

[49] Luo X, Tang R, Long C, Miao Z, Peng Q, Li C. Corrosion behavior of austenitic and ferritic steels in supercritical water. Nuclear Engineering and Technology. 2007;**40**(2):147-154

Pitting Corrosion Resistance and Inhibition of Lean Austenitic Stainless Steel Alloys

Roland Tolulope Loto

Additional information is available at the end of the chapter

http://dx.doi.org/10.5772/intechopen.70579

Abstract

The pitting corrosion behavior of 301, 304 and 316 austenitic stainless steels in 2M H_2SO_4 at 0–1.5% NaCl concentrations was investigated through potentiodynamic polarization and optical microscopy analysis. Electrochemical analysis of the pitting corrosion inhibition and surface protection properties of rosemary oil and aniline on the stainless was also performed. The corrosion rate, pitting potential, passivation potential, metastable pitting potential and surface morphology of both steels where significantly altered by changes in chloride concentration, differences in alloy composition and metallurgical properties of the steels. 316 steel had the lowest corrosion rate and highest pitting corrosion resistance followed by 301 steel. The surface morphology of 316 steel was slightly altered at 1.5% NaCl concentration while 301 steel appears to etch with grain boundaries appearing at higher chloride concentration. 304 steel showed no resistance to pitting after 0% NaCl coupled with relatively significant increase in corrosion rate values. Its surface morphology showed the presence of corrosion pits with respect to chloride and inhibitor concentration. Rosemary oil and aniline significantly reduced the corrosion rates values of the stainless steels and with consequent increase in their pitting corrosion resistance; however the compounds had no positive influence on the pitting corrosion behavior of 304 steel.

Keywords: pitting, corrosion, steel, sodium chloride, austenitic

1. Introduction

Lean austenitic stainless steels consisting mainly of 201, 301, 304 and 316 stainless steel alloys have immense applications in food preparation equipment particularly in chloride environments, boat fittings, chemical containers, heat exchangers, marine applications, pharmaceutical industries, petrochemical, offshore drilling, water desalination and diluted acid containers at low temperature, etc., where high strength is of very high importance in the material of application due to the austenite stability of these alloys which gives a wide range work-hardening rates and ductility [1]. They have less than 20% chromium and 14% nickel content. This category of stainless steels is easily weldable and formable with sufficient corrosion resistance for general purpose applications in mild corrosive environments due to the formation of a passive protective film on the surface of the steels. The passive film consists of the chemical combination of iron and chromium oxy-hydroxide layer, and water containing-compounds at the metal/solution interface [2]. Localized dissolution of the passive film is one of the major causes of pitting corrosion failure of these steels during application. The dissolution tends to be stochastic rather than a visible catastrophic process [3–5].

Previous research on pitting corrosion has shown that it is difficult to experimentally conclude on its mechanism due to its random nature and breakdown of the steel's passive film is due to the selective dissolution of the iron substrate metal [6–12]. The electrochemical process of pitting corrosion results in the formation of microscopic holes of various sizes on metal surfaces which tends to rapidly increase in dimension especially in the presence of chloride ions within the corrosive medium. Under appropriate conditions lean austenitic stainless steels pits at different rates due to differences in chromium, nickel content and metallurgical structure. Due to its localized nature pitting corrosion may be undetected at the onset as only microscopic regions of stainless steel surfaces corrodes while the remaining portion remains passive and cathodic. Pitting is particularly insidious in nature due to the extent of rapidly penetrating into the mass of the metal. The result of such rapid perforations can induce leakage of fluid or alternatively, pitting to crack transition may occur with consequential crack nucleation and growth which leads to brittleness and catastrophic failure. In most cases the corrosive damage is in its advanced stages before detection. The need to further understanding the pitting corrosion mechanism and the influence of organic chemical inhibitors on the electrochemical process resulting in pitting is of very high importance.

1.1. Pitting corrosion mechanism in the presence of chlorides

Pitting corrosion mechanism generally consists of pit initiation and pit propagation stages during the metal dissolution process. The pit initiation stage is the product of the electrochemical action of aggressive ions such as chlorides, sulfates, thiosulfates, etc., at specific regions, sites or flaws in the oxide layer which in some cases causes segregation of alloy elements on the metallic surface. Once a pit nucleates, pit propagation proceeds autocatalytically. The autocatalytic reaction produces cavities initiated at the surface, resulting in a myriad of shapes and sizes. This however depends on the microstructure of the material, electrolyte and various electrochemical factors. Chlorides are most commonly responsible for pit formation on stainless steels as shown in **Figure 1** according to Eqs. (1)–(10). They locally disrupt the passive oxide film at preferential sites especially sites consisting of sulfide inclusions resulting in the gradual formation of corrosion pits. The passive film of stainless steels is made up of adsorbed oxygen. In the presence of chloride ions the higher affinity of oxygen allows for displacement of chloride ions, however as the alloy potential becomes more positive and chloride ions displaces the oxygen atoms and diffuse to the metal/oxide layer [13–20]. The chloride ion diffusion is due to electrostatic attraction. The presence of chloride ions within corrosion pits

http://dx.doi.org/10.5772/intechopen.70579

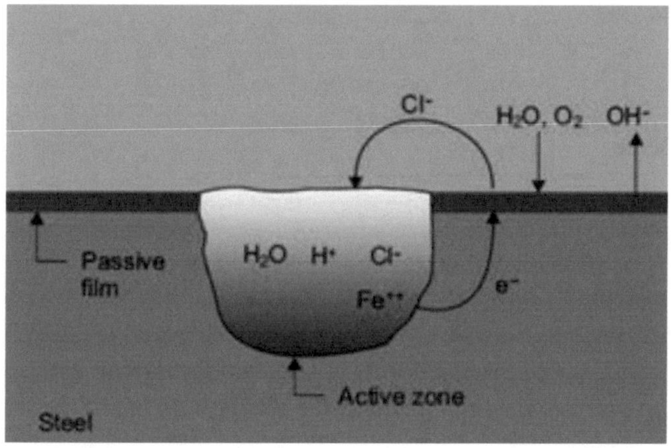

Figure 1. Chloride attack on stainless steel alloy.

stimulates the redox electrochemical reactions necessary for the propagation of the pits. This phenomenon increases the entropy of the reaction species within the pit thus accelerating the localized corrosion of stainless steels. Changes in the electrolyte occurs due to high anodic dissolution rates, and limited diffusion of ionic species The increase in acidity of the electrolyte within the pit caused by insufficient oxygen further accelerates the pitting corrosion reactions in addition to the significant difference between the anode and cathode areas of the steel. This results in metal dissolution leading to cation production within the pits. Pit initiation also occurs due to intermetallic inclusions, micro-segregations, stress points and regions due to dislocations and fatigue.

Anodic reactions inside the pit:

$$3Fe + 4H_2O \rightarrow Fe_3O_4 + 8H^+ + 8e \tag{1}$$

$$Fe \rightarrow Fe^{2+} + 2e^- (\text{dissolution of iron}) \tag{2}$$

$$Fe^{2+} + H_2O \rightarrow Fe(OH)^+ + H^+ \tag{3}$$

$$3Fe(OH)^+ + H_2O \rightarrow Fe_3O_4 + 5H^+ + 2e \tag{4}$$

In the presence of Cl^-, the hydrolysis of Fe^{2+} is accelerated, as shown in the reactions below;

$$Fe^{2+} + Cl^{2-} \rightarrow FeCl_2 \tag{5}$$

$$FeCl_2 + H_2O \rightarrow Fe(OH)^+ + H^+ + 2Cl^- \tag{6}$$

$$Fe^{2+} + H_2O \rightarrow Fe(OH)^+ + H^+ \tag{7}$$

The electrons given up by the anode flow to the cathode (passivated surface) where they are discharged in the cathodic reaction:

$$\frac{1}{2}O_2 + H_2O + 2e^- \leftrightarrow 2(OH^-) \tag{8}$$

$$H^+ + e \rightarrow H \tag{9}$$

$$2H^+ + 2e \rightarrow H_2 \tag{10}$$

Pit propagation involves consistent anodic dissolution by diffusion due to high concentration of metallic and chloride ions within corrosion pits, hence a high concentration of hydrogen ions due to hydrolysis [21]. The corrosion process within the pit is a specific type of anodic reaction whose conditions enables and is necessary for electrochemical reactions within the pit. The diffusion of metallic cations to the pit exterior causes the chloride concentration within the pit to increase resulting in accelerated propagation of the pitting corrosion already taking place. A high cation concentration already exists within the propagating pit, and more chloride ions diffuse into the pit to maintain solution concentration [22, 23]. This phenomenon prevents the repassivation of the stainless steel alloy which otherwise will hinder the pit propagation mechanism.

1.2. Metastable pitting

Repassivation of the oxide protective film on stainless steels causes newly formed pits to disappear for several reasons. Metastable pits are visible during potential scanning indicated by the current fluctuations at very low potentials, below the values necessary for stable pits to occur [24–26]. The pits are microscopic in dimension with a very short lifespan resulting smaller damages on the metallic surface. The passive film on stainless steels is locally damaged during the metastable pitting process before metal dissolution and then surface repassivation. At low potentials, dissolution cannot continue but as the potential of the system increases, the anodic dissolution rate increases and the peak current and lifetime of the metastable pit increase before repassivation. The lower the potential at which stainless steels repassivate after metastable pit formation, the higher the pitting corrosion resistance of the steel. The size of inclusions, flaws, and impurities as well as the presence of fatigue stress are important parameters for the occurrence and properties of metastable pits [27]. There is consistency and correlation between the behavior of stable and metastable pitting. If the acidity and concentration of aggressive ions necessary to keep dissolution in a pit cannot be maintained, repassivation would happen. When a pit has developed to the critical condition which can maintain continuous dissolution in the pit, the metastable pit would transform to a stable pit. The two major criteria responsible for metastable pitting are chloride ion concentration and effect of alloying elements. Chloride in the early stage of pitting corrosion would damage the passive film, and promote the nucleation of metastable pits. The larger the chloride concentration, the higher the metastable pitting nucleation rate. Therefore, with the increase of chloride concentration, both the nucleation and the growth of metastable pits are promoted. According to Burstein [28, 29], this is directly associated with the observation that the number of surface sites available for development of a metastable pit decreased with decreasing chloride for all potentials. A number of researchers studied the effect of alloying elements on the formation of metastable pitting and observed that stainless steel alloys with varying concentration of chromium and molybdenum experience a decrease in the number of metastable events with time due to improve corrosion resistance of the passive film resulting in decrease in available initiation sites [30–33].

2. Potentiodynamic polarization study and inhibition of the pitting corrosion of 301, 304 and 316 in acid chloride media

Pitting corrosion initiation and propagation is subject to factors responsible for any electrochemical corrosion reaction such as charge-transfer mechanisms, ohmic effects and mass transport phenomena. The importance of environmental and material factors relevant to the pitting process, such as electrochemical potential, alloy composition, electrolyte concentration and temperature can be understood by their role on pit growth stability. Influence of the passive film characteristics and the mechanism of the initiation of pitting or breakdown of the otherwise protective passive film are highly important in the study of pitting corrosion. Studies performed on metastable and stable pit growth of metallic alloys have seen considerable progress on propagation processes, conditions and effects of electrochemical variables. Several mechanisms have been proposed to explain the passivity breakdown. Compounds capable of releasing chloride ions to aqueous environments have strong possibility of causing pitting corrosion failure in stainless steels. The chloride ion is highly electronegative and very reactive with specific compounds and elements. The polarization behavior in chloride ion-containing solution has been investigated for decades and a number of conclusions on its electrochemical influence on the pitting corrosion mechanism have been reached [34–41]. In this chapter, the pitting corrosion resistance of type 301, 304 and 316 austenitic stainless steels exposed to 2 M H_2SO_4 acid at specific chloride concentrations will be discussed with focus on the potentiodynamic polarization behavior of the steels, characterization of the pitting susceptibility of the steels in the environments under study, metastable pitting, influence of chloride ion concentration and inhibitor protection through the use of rosemary oil and aniline.

2.1. Experimental methods

2.1.1. Materials and preparation

301, 304, 316 austenitic stainless steels (301SS, 304SS and 316SS) sourced commercially had a nominal composition (wt.%) as shown in **Table 1**. The steel samples machined and afterwards grinded with silicon carbide abrasive papers (80, 120, 220, 320, 600, 800 and 1000 grits) before cleansing with deionized water and acetone for potentiodynamic polarization tests according to ASTM G1–03 [42]. Polarization measurements were conducted out at ambient temperature of 30°C using a three electrode system and glass cell containing 200 mL of the corrosive test solution with Digi-Ivy 2311 potentiostat. 301SS, 304SS, 316SS electrodes mounted in acrylic resin with an exposed surface area of 0.6, 0.79 and 1.33 cm^2 were prepared according to ASTM G59–97 [43]. The polarization plots were obtained at a scan rate of 0.0015 V/s between potentials of −0.5 and +1 V according to ASTM G102–89 [44]. A platinum rod was used as the counter electrode and a silver chloride electrode (Ag/AgCl) as the reference electrode. Corrosion current density (J_{corr}) and corrosion potential (E_{corr}) values were obtained using the Tafel

Element symbol	Si	N	Ni	Mo	Cr	Mn	P	S	C	Fe
% Composition (301SS)	1	0.1	8	–	16	2	0.045	–	0.15	72.7
% Composition (304SS)	0.75	0.1	8	–	18	2	0.045	0.03	0.03	69.31
% Composition (316SS)	0.75	0.1	11	3	18	2	0.045	0.03	0.08	65

Table 1. Percentage nominal composition (wt.%) of 301SS, 304SS and 316SS.

extrapolation method. The corrosion rate (ɤ) and the inhibition efficiency (η_2, %) were calculated from the mathematical relationship:

$$C_R = \frac{0.00327 \times J_{\text{corr}} \times E_{\text{qv}}}{D} \tag{11}$$

where J_{corr} is the current density in A/cm^2, D is the density in g/cm^3 and E_{qv} is the sample equivalent weight in grams. 0.00327 is a constant for corrosion rate calculation in mm/y [45]. 2MH_2SO_4/0.25, 0.5, 0.75, 1, 1.25 and 1.5% NaCl solution, prepared from analar grade of H_2SO_4 acid (98%) and recrystallized NaCl with deionized water. *Rosmarinus officinalis* obtained from NOW Foods, USA is a golden, translucent, oily liquid with a molar mass (active groups) of 691.14 g/mol. Aniline obtained from has a molar mass of 81.38 g/mol. It is a dark translucent liquid soluble in water with a molar mass of 93.13 g/mol. The compounds were prepared in volumetric concentrations of 5% in 200 mL of 2MH_2SO_4/0.25% and 1.5% NaCl solution. Optical images of steel samples before and after corrosion were analyzed with Omax trinocular metallurgical through the aid of ToupCam analytical software.

2.2. Result and discussion

2.2.1. Potentiodynamic polarization studies

The corrosion polarization behavior of 301SS, 304SS and 316SS samples in 2 M H_2SO_4 at 0–1.5% NaCl is shown in **Figures 2–4**. **Table 2** shows the results for the potentiodynamic polarization curves. 316SS generally showed a higher resistance to corrosion than 301SS and 304SS at 0–1.5% NaCl from observation of corrosion rate values in **Table 2**, however its corrosion resistance is subject to changes in chloride ion concentration in the acid solution. The corrosion rates of 304SS were relatively higher than 316SS and 301SS. The surface of 304SS tends to more easily form soft acid, compared to the other steels from the concept Lewis acid–base theory, thus adsorbing chloride and sulfate ions which accelerates its corrosion rate faster than the others [46]. The corrosion rate of 301SS is comparable to 304SS from 0 to 0.5% NaCl having generally similar values but its corrosion rate remained constant after 0.5% due to its more stable electrochemical corrosion resistance behavior to changes in higher chloride ion concentration.

The corrosion rate values of the 301SS, 304SS and 316SS studied corresponds with values of corrosion current and polarization resistance. The cathodic and anodic Tafel slopes for 301SS and 304SS were generally constant with minor variation due to consistent redox electrochemical reactions of hydrogen evolution, oxygen reduction and oxidation reactions taking place at

http://dx.doi.org/10.5772/intechopen.70579

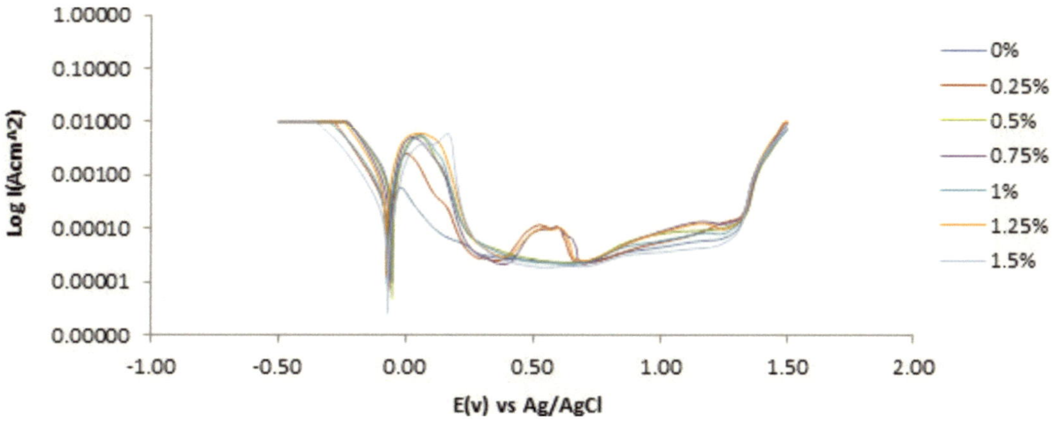

Figure 2. Potentiodynamic polarization curves of 301SS in 2 M H_2SO_4/0–1.5% NaCl.

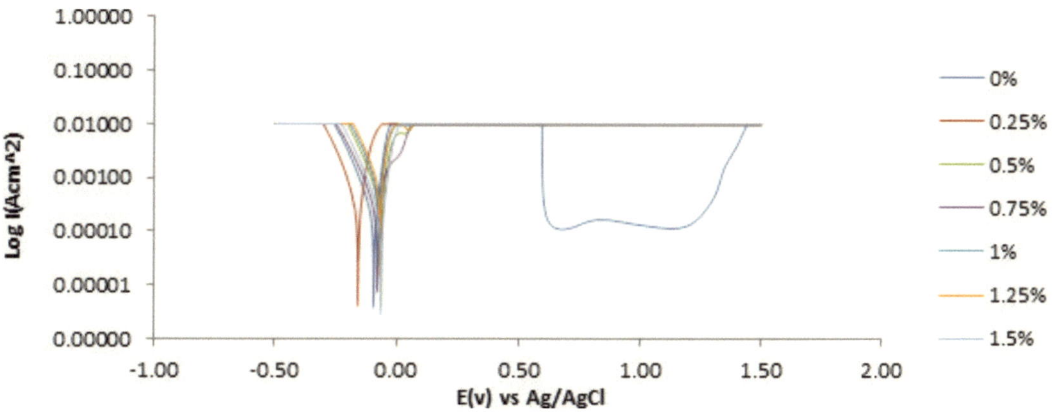

Figure 3. Potentiodynamic polarization curves of 304SS in 2 M H_2SO_4/0–1.5% NaCl.

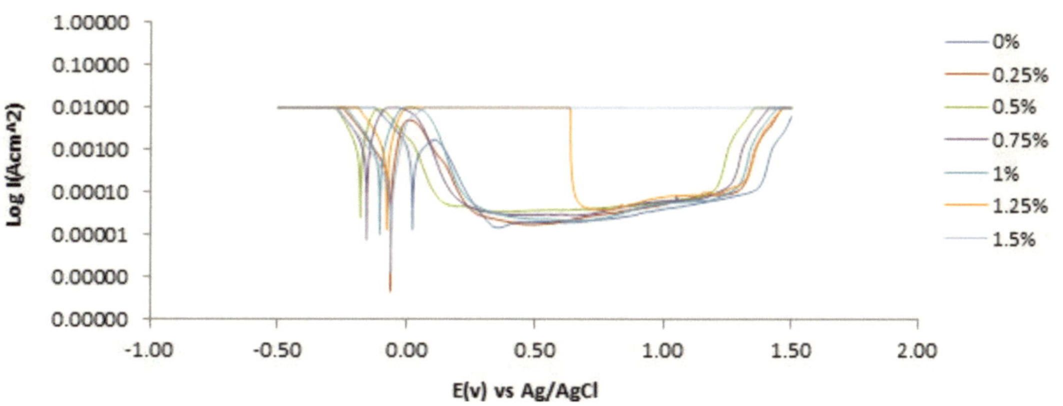

Figure 4. Potentiodynamic polarization curves of 316SS in 2 M H_2SO_4/0–1.5% NaCl.

Sample	2 M H_2SO_4/NaCl conc. (%)	Corrosion rate (mm/y)	Corrosion current (A)	Corrosion current density (A/cm^2)	Corrosion potential (V)	Polarization resistance, R_p (Ω)	Cathodic Tafel slope, B_c (V/dec)	Anodic Tafel slope, B_a (V/dec)
301SS								
A	0	9.086	5.23E-04	8.72E-04	−0.052	31.23	−6.674	3.527
B	0.25	14.889	8.57E-04	1.43E-03	−0.064	16.17	−8.304	3.420
C	0.5	16.314	9.39E-04	1.57E-03	−0.051	14.37	−7.676	3.231
D	0.75	14.681	8.45E-04	1.41E-03	−0.062	19.37	−8.498	3.600
E	1	14.976	8.62E-04	1.44E-03	−0.071	17.39	−7.692	3.513
F	1.25	15.306	8.81E-04	1.47E-03	−0.074	19.17	−8.151	3.620
G	1.5	15.584	8.97E-04	1.50E-03	−0.068	13.11	−8.497	3.030
304SS								
A	0	7.694	5.92E-04	7.49E-04	−0.092	53.57	−10.89	5.289
B	0.25	16.766	1.29E-03	1.63E-03	−0.06	91.83	−11.117	5.433
C	0.5	18.066	1.39E-03	1.76E-03	−0.062	42.35	−10.49	5.146
D	0.75	21.315	1.64E-03	2.08E-03	−0.081	49.01	−10.08	5.127
E	1	22.875	1.76E-03	2.23E-03	−0.073	13.74	−9.026	4.432
F	1.25	26.774	2.06E-03	2.61E-03	−0.067	9.77	−9.853	4.245
G	1.5	30.543	2.35E-03	2.97E-03	−0.062	7.42	−10.91	4.121
316SS								
A	0	1.176	1.50E-04	1.13E-04	0.024	23.34	−6.971	2.486
B	0.25	2.249	2.87E-04	2.16E-04	−0.06	40.34	−8.721	−0.925
C	0.5	3.197	4.08E-04	3.07E-04	−0.178	12.37	−9.826	−1.123
D	0.75	3.879	4.95E-04	3.72E-04	−0.153	13.49	−8.203	2.725
E	1	5.337	6.81E-04	5.12E-04	−0.102	13.92	−8.581	6.976
F	1.25	6.403	8.17E-04	6.14E-04	−0.075	14.49	−8.585	2.033
G	1.5	7.273	9.28E-04	6.98E-04	−0.058	18.29	−8.691	5.979

Table 2. Potentiodynamic polarization results for 301SS, 304SS and 316SS in 2 M H_2SO_4/0–1.5% NaCl.

the steel surfaces with respect to chloride concentration. The observed variation in anodic Tafel slope for 316SS compared to its cathodic Tafel slope with respect to chloride concentration is probably due to the slow electron transfer step resulting from changes in rate controlling step, influence of potential controlled conditions and the presence of molybdenum in its metallurgical structure [47, 48]. This observation corresponds with significant changes in corrosion potential of 316SS compared to 310SS and 304SS. The corrosion potential transits to positive potentials after 0.25% NaCl due to release of fewer electrons which increases the anodic reaction mechanism. The anodic-cathodic polarization scans for 301SS and 304SS (**Figures 2** and 3) were quite similar at all NaCl concentrations compared to 316SS (**Figure 4**) which showed as a wide scatter over the potential domain. This observation shows that changes in NaCl concentration have limited influence on the polarization behavior and redox corrosion reaction mechanisms of 301SS and 304SS as stated earlier from evaluation of anodic and cathodic Tafel slopes.

2.2.2. *Pitting corrosion evaluation*

Study of the pitting corrosion resistance of 301SS, 304SS and 316SS was done through evaluation of pitting, passivation and metastable pitting potentials, and their passivation range. Theoretically the pitting potential is the potential at which pitting corrosion occurs, but below which pits do not nucleate. The passivation potential is the potential value at which potentials greater to or equal to this potential, pit do propagate, but below which the metal retains its passivity. The stainless steel samples exhibited unique pitting corrosion resistance characteristics that differ significantly from each other. 316SS showed the highest resistance to pitting corrosion from observation of potentiostatic values in **Table 3** at 0–1.25% NaCl. At 1.5% NaCl concentration 316SS showed no pitting corrosion resistance behavior from observation of **Figure 3**, as a result the steel failed immediately after anodic polarization. The passivation range of 316SS which shows the extent to which stainless steels sustains their passive film remain unchanged at 0.25% NaCl, after which it increased at 0.5% NaCl and remained

Sample	2 M H_2SO_4/NaCl conc. (%)	Metastable pitting potential (V), E_{mpitt}	Passivation potential (V), E_{pass}	Pitting potential (V), E_{pitt}	Passivation range (V)
310SS					
A	0	−0.07	0.12	1.23	1.11
B	0.25	−0.02	0.23	1.27	1.04
C	0.5	0.01	0.23	1.30	1.07
D	0.75	0.00	0.24	1.29	1.05
E	1	0.02	0.26	1.24	0.98
F	1.25	0.05	0.26	1.24	0.98
G	1.5	0.13	0.28	1.24	0.96
304SS					
A	0	0.55	0.62	1.18	0.56
316SS					
A	0	0.08	0.32	1.33	1.01
B	0.25	−0.02	0.24	1.25	1.01
C	0.5	−0.13	0.13	1.16	1.03
D	0.75	−0.07	0.17	1.20	1.03
E	1	0.00	0.23	1.26	1.03
F	1.25	0.58	0.66	1.29	0.63
G	1.5	0.00	0	0	0

Table 3. Potentiostatic results of pitting passivation and metastable potentials for 301SS, 304SS and 316SS in 2 M H_2SO_4/ 0–1.5% NaCl solution.

constant till 1.25% NaCl at 0.63 V. This observation shows that at a particular chloride concentration 316SS instantaneously loses it passivity. In the presence of chlorides, the pitting potential of 316SS decreased with respect to concentration and remained at values below the pitting potential of the control sample (0% NaCl) due to the action of chloride ions in destroying the passivity of the steel. At 0% NaCl, 316SS passivated at 0.32 V, on addition of chlorides, the steel passivated at lower potentials until 1.25% where it passivated at significantly higher potentials. The phenomenon attests to the prevailing characteristics of the steel in chloride containing environments. A corresponding observation was noted for metastable potential values. Increase in chloride concentration caused a significant rise in the metastable region of the polarization curves of 301SS and 316SS, an indication that the passive film is undergoing localized but transient pitting due to temporary breakdown of the passive film, and the creation and growth of small, occluded cavities before stable passivation. These events are determined by the steels composition and strength of the passive film.

301SS showed generally uniform but narrower passivation behavior over the potentiostatic domain at specific chloride concentrations and retained its passivation behavior at 1.5% NaCl in comparison to 316SS. The passivation range of 310SS decreased after 0% NaCl due to the action of chlorides on the steel, hence its passive protective film reduced in strength compared to 316SS, however the pitting potential values of 301SS increased till 0.75% NaCl before decreasing to values higher than 0% NaCl concentration. This observation do not mean 301SS is more resistant to pitting corrosion than 316SS because in the presence of chlorides 301SS passivates at higher potentials compared to 316SS which passivates at lower potentials hence 316SS has a wider passivation range than 301SS, signifying a more resistance passive film to pitting corrosion. Despite these observations 301SS has better metastable pitting resistance than 316SS as transient pits appears for the steel at higher potentials compared to 316SS. 304SS displayed no pitting corrosion resistance after 0% NaCl concentration. The weak resistance of 304SS is due to excessive adsorption through diffusion of chloride ions at the metal-film interface which induces the electrolytic transport of metallic cations to the acid/chloride solution.

Elemental composition and metallurgical structure are the major factors responsible for the differences in electrochemical behavior and passivation characteristics the three stainless steels studied. Observation of **Table 1** shows that steels have the same elemental composition consisting of an austenite microstructure stabilized by their nickel content, but with the exception of Mo in 316SS. Molybdenum is an important alloy element widely used in metallurgy and has been known to improve the corrosion resistance of stainless steel alloys, being a ferrite former in the presence of manganese and nickel. In the acid/chloride solution it enriches Cr at the metal/solution interface which stabilizes and thickens the passive film [49–57].

2.2.3. Pitting corrosion inhibition

Alloying elements have strong influence on the electrochemical behavior and pitting corrosion resistance of stainless steels. Nickel, chromium, molybdenum and in some cases, titanium, nitrogen, manganese, vanadium are responsible for the properties and strength of the passive film formed on stainless steels [58–60]. Chromium content is one of the major criteria in categorizing austenitic stainless steels and has been observed to significantly change the potentiostatic parameters used in studying pitting corrosion resistance of stainless steels [61, 62]. Sufficient nickel within the iron matrix improves stainless steel resistance to pitting as it stabilizes the austenite phase and limits the ferrite content which if too high may result in lower rust resistance and ductility. Pitting corrosion can also be prevented in many cases with similar methods used to control general corrosion. The most common method of which is the use of chemical compounds to modify the environment during application [63–67]. Studies have shown that inhibitor adsorption on metallic surface depends on the physicochemical characteristics of inhibitor molecule such as functional groups, steric factors, aromaticity, electron density of the donor atoms, π-orbital character of donating electrons, polymerization resulting in formation of protective film and the electronic structure of the molecules [68, 69]. Most of the well-known corrosion inhibitors are organic compounds containing functional groups of heteroatoms capable of chemically reacting through adsorption with valence electrons of stainless steel surfaces at the metal solution interface [70, 71].

In line with the current trend on corrosion inhibition, two known corrosion inhibiting compounds previously used for general corrosion inhibition of carbon steels was used to assess their pitting corrosion potential of the 301SS, 304SS and 316SS under study (ROSO and ANL) [72–74]. The potentiodynamic curve resulting from the use of ROSO compound on 301SS, 304SS and 316SS, and ANL compound on 304SS is shown in **Figure 5**. In the presence of ROSO, the corrosion current densities of 316SS, 304SS and 301SS were significantly influenced, as a result the general corrosion rates of the steel samples at 0.25 and 1.5% NaCl (**Table 4**) reduced drastically due to the electrochemical action of the ROSO molecules in the acid chloride solution. The electrolytic diffusion of chloride and sulfate ions were effectively hindered by the interaction of ROSO molecules with the steel surfaces, which in effect hindered the release of metal cations resulting from anodic oxidation. 316SS had the lowest corrosion rates followed by 301SS. The corrosion potential of 301SS, 304SS and 316SS in the presence of ROSO shifted to negative potentials signifying the dominant cathodic inhibiting property of ROSO [75, 76]. Changes in the cathodic and anodic Tafel slopes were quite similar for 316SS

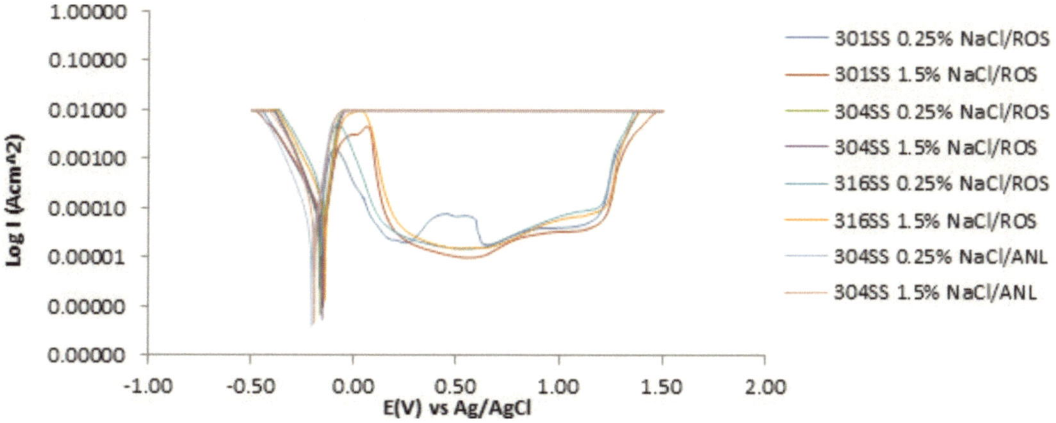

Figure 5. Potentiodynamic polarization curves of 301SS, 304SS and 316SS in 2 M H_2SO_4/5% ROSO at 0.25 and 1.5% NaCl.

Sample	2 M H_2SO_4/ NaCl conc. (%)	Corrosion rate (mm/y)	Corrosion current (A)	Corrosion current density (A/cm^2)	Corrosion potential (V)	Polarization resistance, R_p (Ω)	Cathodic Tafel slope, B_c (V/dec)	Anodic Tafel slope, B_a (V/dec)
ROSO								
301SS	0.25	2.552	1.93E−04	2.45E−04	−0.165	154.67	−8.416	5.107
301SS	1.5	2.684	2.03E−04	2.57E−04	−0.148	140.53	−8.407	5.012
304SS	0.25	8.817	5.15E−04	8.59E−04	−0.161	653.39	−11.790	10.910
304SS	1.5	5.490	3.21E−04	5.35E−04	−0.152	409.85	−11.010	9.230
316SS	0.25	0.234	2.98E−05	2.24E−05	−0.157	1527.33	−7.822	5.604
316SS	1.5	0.565	7.21E−05	5.42E−05	−0.141	1078.19	−8.752	6.950
ANL								
304SS	0.25	0.852	6.56E−05	8.30E−05	−0.204	392.00	−10.710	25.510
304SS	1.5	1.604	1.23E−04	1.56E−04	−0.190	208.20	−11.700	22.470

Table 4. Potentiodynamic polarization results for 301SS, 304SS and 316SS in 2 M H_2SO_4/5% ROSO at 0.25% and 1.5% NaCl.

Sample	2 M H_2SO_4/NaCl Conc. (%)	Metastable pitting potential (V), E_{mpitt}	Passivation potential (V), E_{pass}	Pitting potential (V), E_{pitt}	Passivation range (V)
301SS	0.25%	−0.12	0.13	1.17	1.04
301SS	1.5%	0.03	0.15	1.20	1.05
316SS	0.25%	−0.10	0.11	1.14	1.03
316SS	1.5%	−0.02	0.2	1.20	1.00

Table 5. Potentiostatic results of pitting passivation and metastable potentials for 301SS, 304SS and 316SS in 2 M H_2SO_4/ 5% ROSO at 0.25 and 1.5% NaCl.

and 301SS, but contrast the values obtained for 304SS due to the higher degree of corrosion reactions taking place on 304SS surface. Observation of the potentiostatic data of the three steel samples in **Table 5** showed some mild changes in their values. 316SS at 0.25% NaCl passivated at lower potential of 0.11 V following metastable pitting activity to pit at 1.14 V resulting in a passivation range of 1.03 V while at 1.5% NaCl concentration, pitting corrosion resistance was displayed on the polarization curves resulting in a passivation range of 1.00 V compared to the curve without ROSO compound (**Table 4**, **Figure 4**) where no resistance to pitting corrosion was observed. 301SS passivated at lower potentials following metastable pitting and pitted at higher potentials at 0.25 and 1.5% NaCl concentration resulting in a slightly higher passivation range, thus higher pitting corrosion resistance. ROSO compound had no significant electrochemical influence of the pitting corrosion activity of 304SS despite improved general corrosion resistance. This confirms the earlier statement that metallurgical properties of stainless steel alloys have significant influence on their pitting corrosion resistance. This observation informed the use of ANL compound on 304SS. The corrosion rates of 304SS in the presence of ANL compound significantly decrease further than values obtained in the presence of ROSO

http://dx.doi.org/10.5772/intechopen.70579

compound; however there was no noticeable change in the potentiodynamic polarization curve depicting resistance to pitting corrosion though fewer pits were observed on the steel from optical microscopy analysis which will be discussed later. The observations so far show 304SS has very weak resistance to the electrochemical action of chloride ions. Without chloride 304 displayed limited resistance to pitting corrosion as shown in (**Table 3**, **Figure 3**).

2.2.4. *Thermodynamics of corrosion inhibition*

The adsorption strength of ROSO on 301SS, 304SS and 316SS, and ANL on 304SS was calculated from the thermodynamics of the corrosion inhibition mechanism. Calculated results of Gibbs free energy (ΔG°_{ads}) for the adsorption process is shown in **Table 6**, and evaluated from the mathematical relationship below [77]:

$$\Delta G^{\circ}_{ads} = -2.303RTlog[55.5K_{ads}] \tag{12}$$

where 55.5 is the molar concentration of water in the solution, R is the universal gas constant, T is the absolute temperature and K_{ads} is the equilibrium constant of adsorption. K_{ads} is related to surface coverage (θ) from the Langmuir equation. The presence of impurities, flaws, etc., on studied stainless steel surfaces has a strong influence on results obtained for ΔG°_{ads} [78]. The amount of oxidized metal cations passed into the corrosive media is directly related to the extent of coverage of ROSO and ANL compound. Negative ΔG°_{ads} results show the spontaneity and stability of the adsorption mechanism. Values of ΔG°_{ads} around −20 kJ/mol shows physisorption adsorption reaction while values around −40 kJ/mol or higher involve charge sharing or chemisorption due to chemical interaction among the reacting species [79]. The ΔG°_{ads} values obtained for ROSO and ANL interaction on the 301SS, 304SS and 316SS shows chemisorption adsorption of ROSO and ANL molecules on the steel surfaces in response to competitive adsorption of chloride and sulfate ions.

Specimen	NaCl concentration (%)	Surface coverage (θ)	Equilibrium constant of adsorption (K)	Gibbs free energy, ΔG ($kjmol^{-1}$)
ROSO				
301SS	0.25	0.829	66825.7	−37.48
301SS	1.5	0.828	66439.8	−37.47
304SS	0.25	0.474	12463.8	−33.32
304SS	1.5	0.820	63082.0	−37.34
316SS	0.25	0.896	119302.8	−38.92
316SS	1.5	0.922	164090.6	−39.71
ANL				
304SS	0.25	0.949	34792.7	−35.87
304SS	1.5	0.947	33608.3	−35.78

Table 6. Data for Gibbs free energy (ΔG°_{ads}), surface coverage (θ) and equilibrium constant of adsorption (K_{ads}) for ROSO and ANL adsorption on 301SS, 304SS and 316SS.

2.2.5. Optical microscopy analysis of 301SS, 304SS and 316SS morphology

The optical microscopy images of 301SS, 304SS and 316SS before corrosion and after the corrosion test at 0, 0.25 and 1.5% NaCl are shown from **Figures 6(a)–8(c)** at mag. 40×. **Figure 9 (a)–Figure 11(b)** shows the images of 301SS, 304SS and 316SS after corrosion in the presence of ROSO compound while **Figure 11(a, b)** shows the images of 304SS after corrosion in the presence of ANL compound. Severe morphological deterioration is clearly visible on 304SS (**Figure 7(a–d)**) due to the action of chloride and sulfate ions. At 0% NaCl (**Figure 7(b)**) the serrated edges and lines on the steel surface are faintly visible due to corrosion. Significant number of corrosion pits can be observed due to surface oxidation and release of metal cations into the solution, though sulfate ions is solely responsible for these observations, however at 0.25% NaCl (**Figure 7(c)**) the number of corrosion pits have increased significantly and they appear to be deeper due to the action of chloride ions. The corrosion pits in **Figure 7(d)** appear to be smaller, while the surface morphology seems rougher then the image in **Figure 7(c)**. Results from potentiodynamic study shows the highest corrosion rate for 304SS at 1.5% NaCl, thus it is suggested that the presence of excessive chloride ions in solution does not necessarily

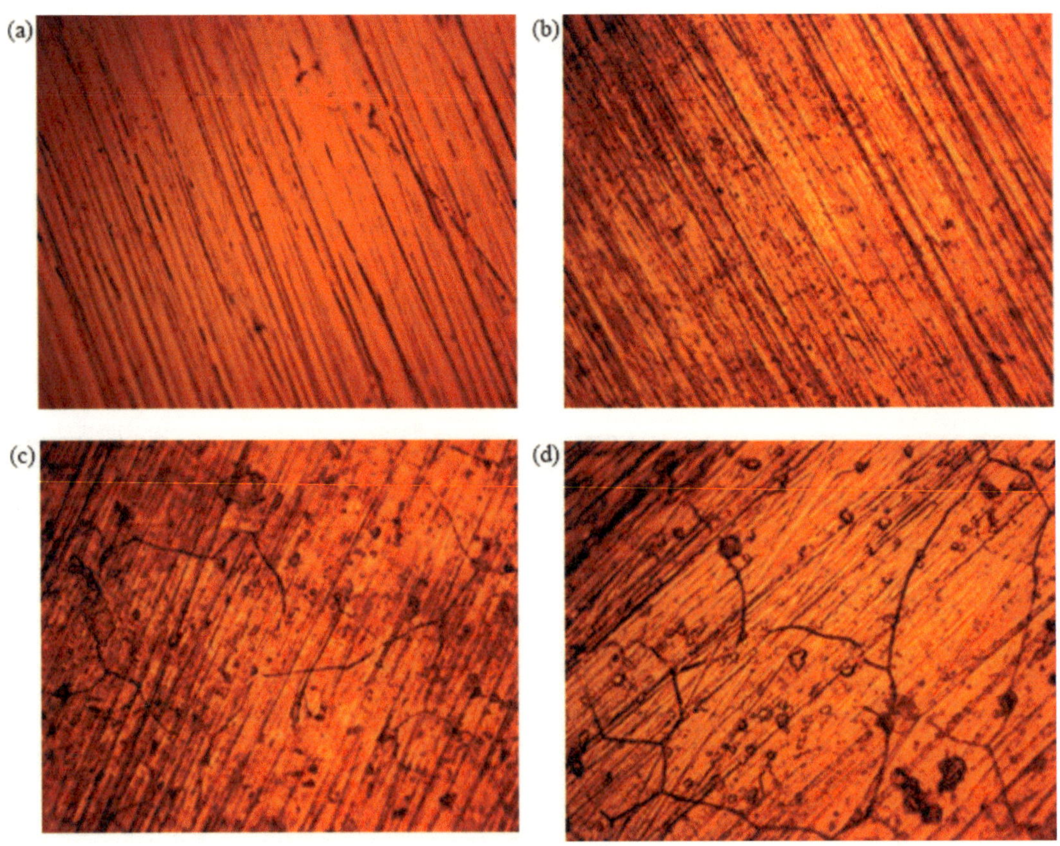

Figure 6. Optical microscopy image 301SS at mag. 40× (a) before corrosion, (b) at 0% NaCl, (c) at 0.25% NaCl and (d) at 1.5% NaCl.

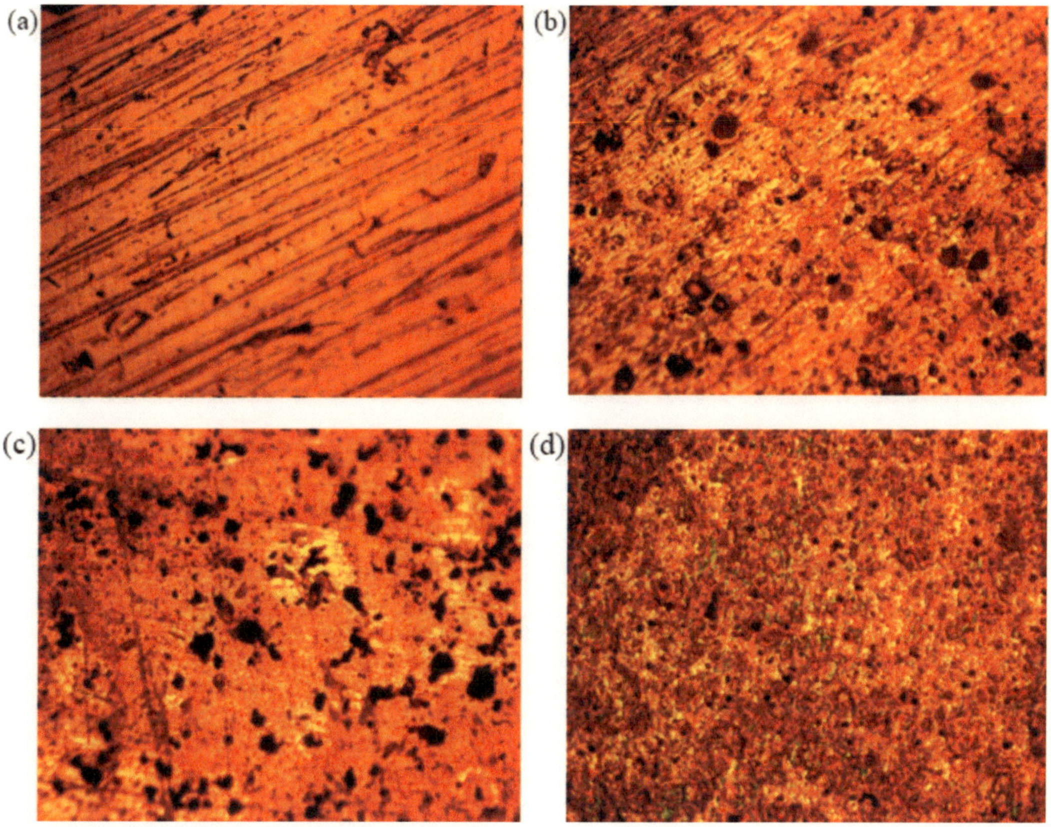

Figure 7. Optical microscopy image 304SS at mag. 40× (a) before corrosion, (b) at 0% NaCl, (c) at 0.25% NaCl and (d) at 1.5% NaCl.

mean more corrosion pits even though the general corrosion rate may be higher. There seems to be a threshold level of chloride concentration responsible for the size of corrosion pits.

Corrosion pits are clearly absent from 301SS (**Figure 7(b–d)**) due to the resilience of its passive film. What appears on its morphology (**Figure 7(c)**) seems to be shallow indentations and faint appearance of the grain boundaries due to mild etching by the chloride ions. **Figure 7(d)** shows a worn-out morphology compared to **Figure 7(c)** with the indentation larger and the grain boundary much more visible. In general, the morphology of 301SS shows a highly resistant steel to pitting. The morphology of 316SS remained unchanged or probably etched at 0% NaCl (**Figure 8(b)**). In 0.25% NaCl, (**Figure 8(c)**) the surface morphology seems to have worn out compared to **Figure 8(b)**. These observations show the electrochemical action of chlorides and sulfates have limited influence on the pitting corrosion resistance of 316SS. At 1.5% NaCl (**Figure 8(d)**), significantly morphological deterioration occurred; the grain boundaries are also faintly visible with numerous micro indentations. The morphologies of 301SS and 316SS at 1.5% NaCl (**Figures 9(b)** and **11(b)**) in the presence of ROSO compound remained generally the same even though there was significant improvement in the corrosion rate values from potentiodynamic polarization test, however there seems to be mild improvement for the

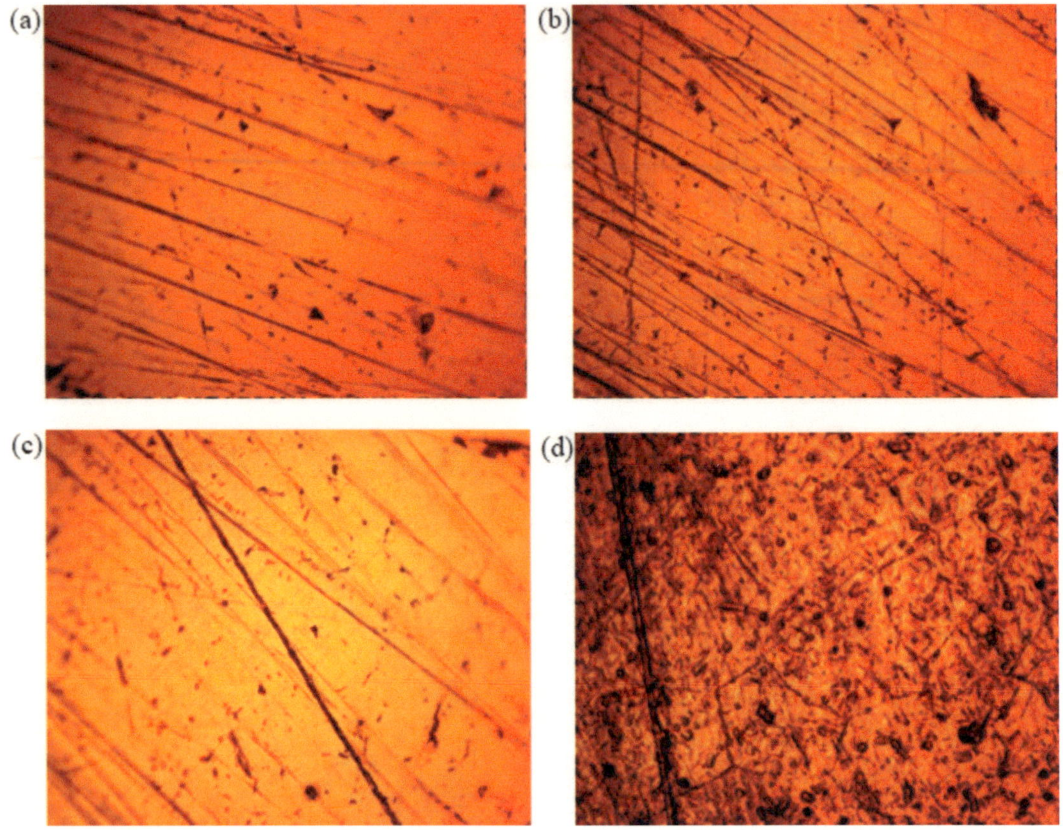

Figure 8. Optical microscopy image 316SS at mag. 40× (a) before corrosion, (b) at 0% NaCl, (c) at 0.25% NaCl and (d) at 1.5% NaCl.

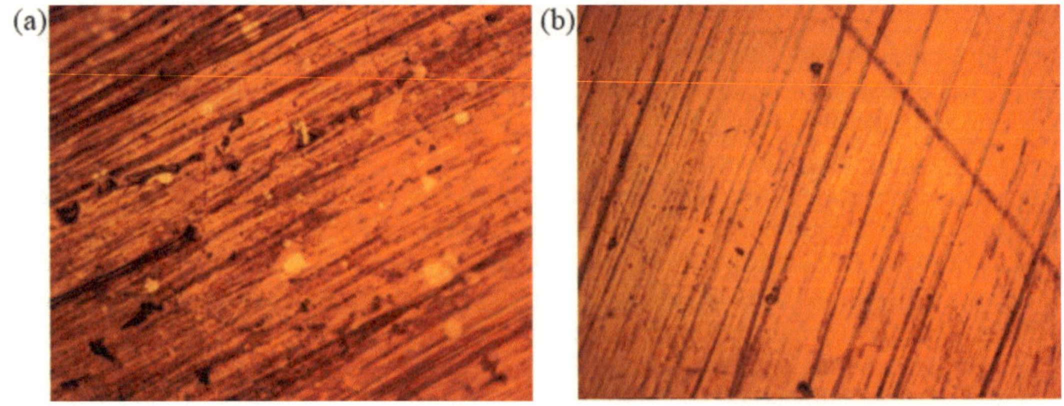

Figure 9. Optical microscopy image 301SS in the presence of ROSO compound (a) at 0.25% NaCl and (b) at 1.5% NaCl.

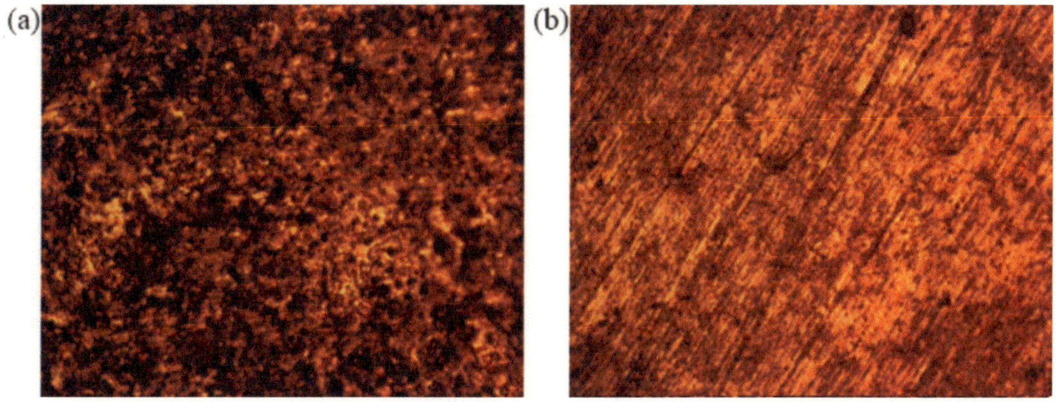

Figure 10. Optical microscopy image 304SS in the presence of ROSO compound (a) at 0.25% NaCl and (b) at 1.5% NaCl.

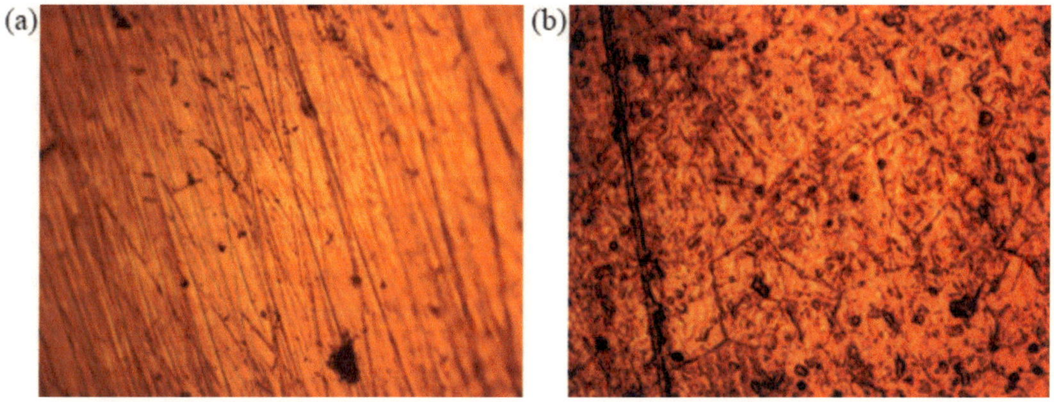

Figure 11. Optical microscopy image 316SS in the presence of ROSO compound (a) at 0.25% NaCl and (b) at 1.5% NaCl.

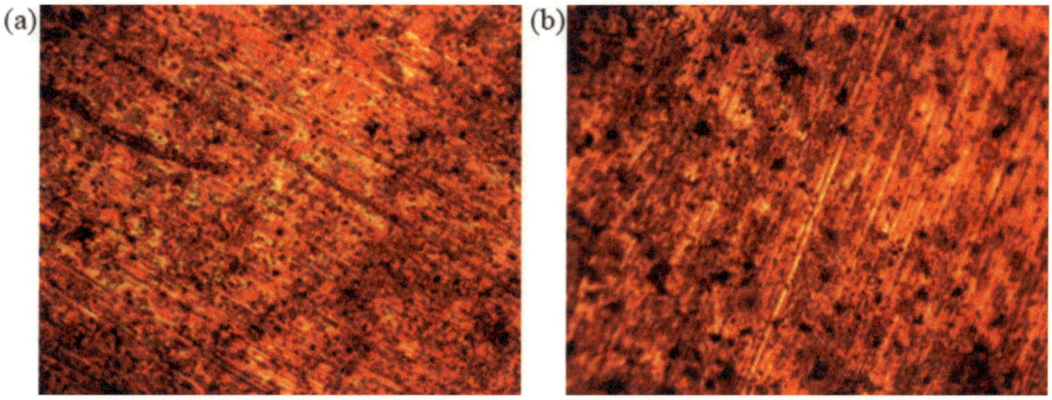

Figure 12. Optical microscopy image 304SS in the presence of ANL compound (a) at 0.25% NaCl and (b) at 1.5% NaCl.

images in Figures **9(a)** and **11(a)**. 304SS image (**Figure 10(a)**) showed significant increase in the number of micro-pits despite improvement in general corrosion rate value from the corrosion test. There appears to be smaller but more micro-pits compared to fewer but larger pits in **Figure 7(c)**, while **Figure 10(b)** showed no visible pits compared to **Figure 7(d)** in the presence of ROSO. Comparing the observation on 304SS with **Figure 12(a)** and (b), the optical images of 304SS from the corrosive solution in the presence of ANL compound are quite different from the images obtained in the presence of ROSO. There seems to be a remarkable improvement in the morphology of 304SS at 0.25%NaCl/ANL compared to 304SS at 0.25%NaCl/ROSO. The corrosion pits are significantly smaller and the steel surface is not as badly damaged like the surface from ROSO compound. Quite the contrary observation was obtained for 304SS at 1.5% NaCl/ANL compared to 304SS at 1.5%NaCl/ROSO. These observations show that chemical compounds are specific in action during corrosion inhibition.

3. Conclusion

Chloride concentration in 2 M H_2SO_4 had strong electrochemical effect on the pitting corrosion resistance and passivation behavior 301SS, 304SS and 316SS. 316SS showed the highest pitting corrosion resistance with resilient passivation behavior while 304SS failed immediately after anodic polarization in the presence of chlorides. The corrosion rates of the stainless steels generally increased with increase in chloride concentration. ROSO compound significantly reduced the corrosion rates of the stainless steels studied with notable improvement in their potentiodynamic polarization and passivation behavior; however it had detrimental effect on the pitting susceptibility of 304SS at lower chloride concentrations. ANL compound further reduced the corrosion rates of the stainless steels with particular improvement in morphology of 304SS but no changes in the potentiodynamic polarization and passivation behavior of 304SS was observed, hence limited pitting corrosion inhibition.

Acknowledgements

The author is grateful to Covenant University, Ota, Ogun State, Nigeria for the funding of the research and provision of research facilities.

Author details

Roland Tolulope Loto

Address all correspondence to: tolu.loto@gmail.com

Department of Mechanical Engineering, Covenant University, Ota, Ogun State, Nigeria

References

[1] Alves H, Agarwal DC, Werner H. Nace–International Corrosion Conference Series; 2006, Houston, Texas.

[2] Pardo A, Merino MC, Coy AE, Viejo F, Arrabal R, Matykina E. Effect of Mo and Mn additions on the corrosion behaviour of AISI 304 and 316 stainless steels in H_2SO_4. Corrosion Science. 2008;**50**(3):780-794. DOI: 10.1016/j.corsci.2007.11.004

[3] Refaey SAM, Taha F, Abd El-Malak AM. Corrosion and inhibition of 316L stainless steel in neutral medium by 2-mercaptobenzimidazole. International Journal of Electrochemical Science. 2006;**1**:80-91

[4] Williams DE, Westcott C, Fleischmann M. Stochastic models of pitting corrosion of stainless steels. II. Measurement and interpretation of data at constant potential. Journal of the Electrochemical Society. 1985;**132**:1796-1804. DOI: 10.1149/1.2114221

[5] Shibata T, Takeyama T. Pitting corrosion as a stochastic process. Nature. 1976;**260**:315-316. DOI: 10.1038/260315a0

[6] Szklarska-Smialowska Z. Progress in understanding pitting corrosion. Electrochemical Society Proceedings. 2002;**13**:251-265

[7] Soltis J. Passivity breakdown, pit initiation and propagation of pits in metallic materials. Corrosion Science. 2015;**90**:5-22. DOI: 10.1016/j.corsci.2014.10.006.

[8] Newman RC, Foong TM, Sieradzki K. Validation of a percolation model for passivation of Fe.Cr alloys: I current efficiency in the incompletely passivated state. Corrosion Science. 1988;**28**(5):523−527. DOI: 10.1016/0010-938X(88)90074-1

[9] Song Q, Newman RC, Cottis RA, Sieradzki K. Validation of a percolation model for passivation of Fe–Cr alloys: Two-dimensional computer simulations. Journal of the Electrochemical Society. 1990;**137**:435-439. DOI: 10.1149/1.2086458

[10] Song Q, Newman RC, Cottis RA, Sieradzki K. Computer simulation of alloy passivation and activation. Corrosion Science. 1990;**31**:621−626. DOI: 10.1016/0010-938X(90)90171-Z

[11] Burstein GT, Marshall PI. The coupled kinetics of film growth and dissolution of stainless steel repassivating in acid solutions. Corrosion Science. 1984;**24**(5):449−462. DOI: 10.1016/0010-938X(84)90070-2

[12] Uhlig HH. The absorption theory of passivity and the flade potential. Zeitscrift für Elektrochemie. 1958;**62**(6–7):626-632

[13] Macdonald DD. The point defect model for the passive state. Journal of the Electrochemical Society. 1992;**139**(12):3434-3449

[14] Ahn SJ, Won HSK, Macdonald DD. Role of chloride ion in passivity breakdown on iron and nickel. Journal of the Electrochemical Society. 2005;**152**(11):B482-B490. DOI: 10.1149/1.2048247

[15] Kazuhiro Teramoto K, Asami K, Hashimoto K. The composition of passive films on ferritic 30 Cr stainless steels in H_2SO_4. Boshoku Gijutsu. 1978;**27**:57-61

[16] Hoar T, Mears D, Rothwell G. The relationships between anodic passivity, brightening and pitting. Corrosion Science. 1965;**5**(4):279-289. DOI: 10.1016/S0010-938X(65)90614-1

[17] Streicher MA. Pitting corrosion of 18Cr-8Ni stainless steel. 1956;**103**(7):375–390. DOI: 10.1149/1.2430359

[18] Matsuda S, Uhlig H. Effect of pH, sulfates, and chlorides on behavior of sodium chromate and nitrite as passivators for steel. Journal of the Electrochemical Society. 1964;**111**(2):156-161

[19] Uhlig HH, Lord SS Jr. Amount of oxygen on the surface of passive stainless steel. Journal of the Electrochemical Society. 1953;**100**(5):216-221. DOI: 10.1149/1.2781107

[20] Kolotyrkin JM. Effects of anions on the dissolution kinetics of metals. Journal of the Electrochemical Society. 1961;**108**(3):209-216. DOI: 10.1149/1.2428048

[21] Eklund GS. The effect of additives on the reaction mechanism of the $Pb/PbSO_4$ electrode. Journal of the Electrochemical Society. 1974;**121**(4):467-473. DOI: 10.1149/1.2401839

[22] Pistorius PC, Burstein GT. Metastable pitting corrosion of stainless steel and the transition to stability. Philosophical Transactions of the Royal Society A. 1992;**341**:531-542. DOI: 10.1098/rsta.1992.0114

[23] Burstein GT, Mattin SP. Repetitive nucleation of corrosion pits on stainless steel and the effects of surface roughness. Journal of the Electrochemical Society. 2001;**148**(12): B504-B516. DOI: 10.1149/1.1416503

[24] Frankel GS. Pitting corrosion of metals a review of the critical factors. Journal of the Electrochemical Society. 1998;**145**(6):2186-2198

[25] Pride ST, Scully JR, Hudson JL. Metastable pitting of aluminum and criteria for the transition to stable pit growth. Journal of the Electrochemical Society. 1994;**141**(11):3028-3040. DOI: 10.1149/1.2059275

[26] Laycock NJ, Newman RC. Localised dissolution kinetics, salt films and pitting potentials. Corrosion Science. 1997;**39**(10–11):1771-1790. DOI: 10.1016/S0010-938X(97)00049-8

[27] Suter T, Webb E, Böhni H, Alkire RC. Pit initiation on stainless steels in 1 M NaCl with and without mechanical stress. Journal of the Electrochemical Society. 2001;**148**(5):B174-B185. DOI: 10.1149/1.2059275

[28] Pistorius PC, Burstein GT. Aspects of the effects of electrolyte composition on the occurence of metastable pitting on stainless steel. Corrosion Science. 1994;**36**:525-538. DOI: 10.1016/0010-938X(94)90041-8

[29] Burstein GT, Mattin SP. Nucleation of corrosion pits on stainless steel. Philosophical Magazine Letters. 1992;**66**(1–4):127-131. DOI: 10.1016/0010-938X(93)90133-2

[30] Burstein GT, Pistorius PC, Mattin SP. The nucleation and growth of corrosion pits on stainless steel. Corrosion Science. 1993;**35**(1–4):57-62. DOI: 10.1016/0010-938X(93)90133-2

[31] Williams DE, Stewart J, Balkwill PH. The nucleation, growth and stability of micropits in stainless steel. Corrosion Science. 1994;**36**(7):1213-1235. DOI: 10.1016/0010-938X(94)90145-7

[32] Virtanen S, Kobayashi Y, Böhni H. In: Kelly RG, Frankel GS, Natishan PM, Newman RC, editors. Critical Factors in Localized Corrosion III. Pennington: Electrochemical Society; 1999. p. 281-290

[33] Ilevbare GO, Burstein GT. The role of alloyed molybdenum in the inhibition of pitting corrosion in stainless steels. Corrosion Science. 2001;**43**(3):485-513. DOI: 10.1016/S0010938X(00)00086-X

[34] Wu J, Li X, Du C, Wang S, Song Y. Effects of Cl^- and SO_4^{2-} ions on corrosion behavior of X70 steel. Journal of Materials Science and Technology. 2005;**21**(1):28-32

[35] Ericson S, Kurol J. Inhibitive action of cannabis plant extract on the corrosion of copper in 0.5M H_2SO_4. International Journal of Electrochemical Science. 2013;**8**:5851-5865

[36] Cabral-Miramontes JA, Barceinas-Sánchez JDO, Poblano-Salas CA, Pedraza-Basulto GK, Nieves-Mendoza D, Zambrano Robledo PC, Almeraya-Calderón F, Chacón-Nava JG. Corrosion behavior of AISI 409Nb stainless steel manufactured by powder metallurgy exposed in H_2SO_4 and NaCl solutions. International Journal of Electrochemical Science. 2013;**8**:564-577

[37] Saadawy M. Effect of inorganic anions on the pitting behaviour of austenitic stainless steel 304 in H_2SO_4 solution containing chloride ion. International Journal of Electrochemical Science. 2016;**11**:2345-2359

[38] Loto RT. Pitting corrosion evaluation of austenitic stainless steel type 304 in acid chloride media. Journal of Materials and Environmental Science. 2013;**4**:448-459

[39] Yi Y, Cho P, Al Zaabi A, Addad Y, Jang C. Potentiodynamic polarization behaviour of AISI type 316 stainless steel in NaCl solution. Corrosion Science. 2013;**74**:92-97. DOI: 10.1016/j.corsci.2013.04.028

[40] Guan L, Zhang B, Yong XP, Wang JQ, Han EH, Ke W. Effects of cyclic stress on the metastable pitting characteristic for 304 stainless steel under potentiostatic polarization. Corrosion Science. 2015;**93**:80. DOI: 10.1016/j.corsci.2015.01.009

[41] Réguer S, Dillmann P, Mirambet F. Buried iron archaeological artefacts: Corrosion mechanisms related to the presence of Cl^- containing phases. Corrosion Science. 2007;**49**(6): 2726-2744. DOI: 10.1016/j.corsci.2006.11.009

[42] ASTM G1–03. Standard Practice for Preparing, Cleaning, and Evaluating Corrosion Test Specimens [Internet]. 2011. Available from: http://www.astm.org/Standards/G1 [Accessed: May 30, 2016].

[43] ASTM G59–97. Standard Test Method for Conducting Potentiodynamic Polarization Resistance Measurements [Internet]. 2014. http://www.astm.org/Standards/G31 [Accessed: May 30, 2016].

[44] ASTM G102–89. Standard Practice for Calculation of Corrosion Rates and Related Information from Electrochemical Measurements [Internet]. 2015. e1. http://www.astm.org/Standards/G31 [Accessed: May 30, 2016].

[45] Choi Y, Nesic S, Ling S. Effect of H_2S on the CO_2 corrosion of carbon steel in acidic solutions. Electrochimica Acta. 2011;**56**(4):1752-1760. DOI: 10.1016/j.electacta.2010.08.049

[46] Jensen WB. The Lewis Acid-Base Concepts. New York: John Wiley and Sons, Inc.; 1980. p. 112-336. DOI: 10.1002/nadc.19800281014

[47] Bockris JOM, Drazic D, Despic AR. The electrode kinetics of the deposition and dissolution of iron. Electrochimica Acta. 1961;**4**(2–4):325-361. DOI: 10.1016/0013-4686(61)80026-1

[48] Bockris JOM, Kita H. The temperature coefficients of electrode potentials: II. The second isothermal temperature coefficient. Journal of the Electrochemical Society. 1961;**108**(7): 676-685. DOI: 10.1149/1.2428187

[49] International Stainless Steel Forum [Internet]. http://www.worldstainless.org/Files/issf/non-image-files/PDF/TheStainlessSteelFamily.pdf. [Accessed: March 30, 2017].

[50] Metallurgy of Mo in Stainless Steel, International Molybdenum Association [Internet]. http://www.imoa.info/molybdenum-uses/molybdenum-grade-stainless-steels/metallurgy-of-molybdenum-in-stainless-steel.php. [Accessed: March 30, 2017].

[51] Hiramatsu N, Stott FH. The effects of molybdenum on the high-temperature oxidation resistance of thin foils of Fe–20Cr–5Al at very high temperatures. Oxidation of Metals. 2000;**53**(56):561-576. DOI: 10.1023/A:1004689211302

[52] Abreu CM, Cristobal MJ, Losada R, Novoa XR, Pena G, Perez MC. Comparative study of passive films of different stainless steels developed on alkaline medium. Electrochimica Acta. 2004;**49**(17–18):3049-3056. DOI: 10.1016/j.electacta.2004.01.064

[53] Montemor MF, Simoes A, Ferreira MGS, Da Cunha MB. The role of Mo in the chemical composition and semiconductive behaviour of oxide films formed on stainless steels. Corrosion Science. 1999;**41**(1):17-34. DOI: 10.1016/S0010-938X(98)00126-7

[54] Lu YC, Clayton CR. An XPS study of the passive and transpassive behavior of molybdenum in deaerated 0.1 M HCl. Corrosion Science. 1989;**29**(8):927-937. DOI: 10.1016/0010-938X(89)90085-1

[55] Vignal V, Olive JM, Desjardins D. Effect of molybdenum on passivity of stainless steels in chloride media using ex situ near field microscopy observations. Corrosion Science. 1999;**41**(5):869-884

[56] Olefjord I, Brox B, Jelvestam U. Surface composition of stainless steels during anodic dissolution and passivation studied by ESCA. Journal of the Electrochemical Society. 1985;**132**(12):2854-2861. DOI: 10.1149/1.2113683

[57] Schultze JW, Lohrengel MM, Ross D. Nucleation and growth of anodic oxide films. Electrochimica Acta. 1983;**28**(7):973-984

[58] Sedriks AJ. Corrosion of Stainless Steels. New York, NY: John Wiley and Sons; 1996

[59] Schmuki P, Virtanen S, Isaacs HS, Ryan MP, Davenport AJ, Bohni H, Stenberg T. Electrochemical behavior of Cr_2O_3/Fe_2O_3 artificial passive films studied by in situ XANES. Journal of the Electrochemical Society. 1998;**145**(3):791-801. DOI: 10.1149/1.1838347

[60] Newman RC. The dissolution and passivation kinetics of stainless alloys containing molybdenum—1. Coulometric studies of Fe.Cr and Fe.Cr.Mo alloys. Corrosion Science. 1985;**25**(5):331-339. DOI: 10.1016/0010-938X(85)90111-8

[61] Szklarska-Smialowska Z. Pitting Corrosion of Metals. Houston, TX: NACE; 1986. p. 430

[62] Horvath J, Uhlig HH. Critical potentials for pitting corrosion of Ni, Cr-Ni, Cr-Fe, and related stainless steels. Journal of the Electrochemical Society. 1968;**115**(8):791-795. DOI: 10.1149/1.2411433

[63] Ochao N, Moran F, Pebre N. The synergistic effect between phosphonocarboxylic acid salts and fatty amines for the corrosion protection of a carbon steel. Journal of Applied Electrochemistry. 2004;**34**(5):487-493

[64] Oguzie EE. Influence of halide ions on the inhibitive effect of congo red dye on the corrosion of mild steel in sulphuric acid solution. Materials Chemistry and Physics. 2004;**87**(1):212-217. DOI: 10.1016/j.matchemphys.2004.06.006

[65] Shalaby MN, Osman MM. Synergistic inhibition of anionic and nonionic surfactants on corrosion of mild steel in acidic solution. Anti-Corrosion Methods and Materials. 2001;**48**(5): 309-318. DOI: 10.1108/EUM0000000005885

[66] Ebenso EE. Synergistic effect of halide ions on the corrosion inhibition of aluminium in H_2 SO_4 using 2-acetylphenothiazine. Materials Chemistry and Physics. 2003;**79**(1):58-70. DOI: 10.1016/S0254-0584(02)00446-7

[67] Jayalakshmi M, Muralidharan VS. Correlation between structure and inhibition of organic compounds for acid corrosion of transition metals. Indian Journal of Chemical Technology. 1998;**5**:16-28

[68] Granese SL, Rosales BM, Oviedo C, Zebrino JO. The inhibition action of heterocyclic nitrogen organic compounds on Fe and steel in HCl media. Corrosion Science. 1992;**33**(9): 1439-1453. DOI: 10.1016/0010-938X(92)90182-3

[69] Raicheva SN, Aleksiev BV, Sokolova EI. The effect of the chemical structure of some nitrogen- and sulphur-containing organic compounds on their corrosion inhibiting action. Corrosion Science. 1993;**34**(2):343-350. DOI: 10.1016/0010-938X(93)90011-5

[70] Abd El Rehim SS, Ibrahim MAM, Khaled KF. 4-Aminoantipyrine as an inhibitor of mild steel corrosion in HCl solution. Journal of Applied Electrochemistry. 1999;**29**(5):593-599

[71] Behpour M, Ghoreishi SM, Soltani N, Salavati-Niasari M, Hamadanian M, Gandomi A. Electrochemical and theoretical investigation on the corrosion inhibition of mild steel by

thiosalicylaldehyde derivatives in hydrochloric acid solution. Corrosion Science. 2008;**50**(8): 2172-2188. DOI: 10.1016/j.corsci.2008.06.020

[72] Loto RT, Oghenerukewe E. Inhibition studies of rosmarinus officinalis on the pitting corrosion resistance 439LL ferritic stainless steel in dilute sulphuric acid. Oriental Journal of Chemistry. 2016;**32**(5):2813-2832. DOI: 10.13005/ojc/320557

[73] Loto RT, Loto RO, Joseph OO, Akinwumi I. Electrochemical studies of the corrosion inhibition property of rosmarinus officinalis on mild steel in dilute sulphuric acid. Journal of Chemical and Pharmaceutical Research. 2015;**7**(7):105-116

[74] Bendahou M, Benabdellah M, Hammouti B. A study of rosemary oil as a green corrosion inhibitor for steel in 2M H_3PO_4. Pigment & Resin Technology. 2006;**35**(2):95-100. DOI: 10.1108/03699420610652386

[75] Susai RS, Mary R, Noreen A, Ramaraj R. Synergistic corrosion inhibition by the sodium dodecylsulphate–Zn^{2+} system. Corrosion Science. 2002;**44**(11):2243-2252. DOI: 10.1016/S0010-938X(02)00052-5

[76] Sahin M, Bilgiç S, Yılmaz H. The inhibition effects of some cyclic nitrogen compounds on the corrosion of the steel in NaCl mediums. Applied Surface Science. 2002;**195**(1–4):1-7. DOI: 10.1016/S0169-4332(01)00783-8

[77] Aharoni C, Ungarish M. Kinetics of activated chemisorption. Part 2. Theoretical models. Journal of the Chemical Society, Faraday Transactions. 1977;**73**:456-464

[78] Lowmunkhong P, Ungthararak D, Sutthivaiyakit P. Tryptamine as a corrosion inhibitor of mild steel in hydrochloric acid solution. Corrosion Science. 2010;**52**(1):30-36. DOI: 10.1016/j.corsci.2009.08.039

[79] Vračar LM, Dražić DM. Adsorption and corrosion inhibitive properties of some organic molecules on iron electrode in sulfuric acid. Corrosion Science. 2002;**44**(8):1669-1680. DOI: 10.1016/S0010-938X(01)00166-4

Chapter 9

Friction Welding of Austenitic Stainless Steel with Copper Material

Shanjeevi Chinnakannan

Additional information is available at the end of the chapter

http://dx.doi.org/10.5772/intechopen.70558

Abstract

Austenitic stainless steels are most preferred over other types of stainless steel families. Welding of stainless steel using friction welding is widely seen in the current scenario. Since the time consumed for friction welding is very less, metallurgical defects are almost reduced without pre- and postheat treatment. The problems encountered in friction welding during joining of austenitic stainless steel are very limited when compared to fusion welding process. The studies have undergone with joining of austenitic stainless steel and copper material to evaluate the friction welding parameter for finding the good bond strength.

Keywords: friction welding, AFM, copper, stainless steel, microstructure

1. Introduction to stainless steel

Among the various sources of materials available, selection of stainless steel is one of the important classes of engineering materials considered in the past and present scenario. According to chemical elements, the stainless steel is classified into different grades with respect to microstructures such as ferritic, austenitic, martensitic and duplex stainless steel (the combination of austenite and ferrite). These different grades have been used in various applications. The most common uses are listed below:

- Automotive and transportation
- Architecture and construction
- Food and catering

- Medical
- Energy and heavy industries

It is the primary stainless steel used in aviation construction. The grades with 3xx series are often referred as austenitic stainless steel. Each grade has followed with a specified letter that represents chemical element information. Low carbon austenitic stainless steel is represented with 'L'; High carbon steel with 'H'; Nitrogen bearing steel with 'N'; some cases with modified composition say 'LN' from base alloy.

2. Friction welding and its importance

Friction welding is a metal-joining process made by continuous-rubbing action at the interface of two different materials, which leads to heat dissipation. Due to continuous action of rotation, the heat generated at the interface results in deformation to the plastic stage by the conversion of mechanical energy into thermal energy under pressure resulting in good bond strength of the material.

Friction welding is more economical and time-consuming, which requires a low input of energy and high production rate with less material wastage in joining dissimilar metals or alloys. During friction welding of steels, the weld interface produces heat with temperature range of 900 to 1300°C.

3. Problems in fusion welding on stainless steel

In general, austenitic stainless steels are easily weldable [1]. Based on physical properties on ferritic, martensitic and duplex stainless steels, austenitic stainless steel is considerably different than others [2]. In fusion welding process, particularly in gas tungsten, electron beam and laser welding, there is a possibility with unexpected phase propagation. Due to metallurgical changes in weld interface, phase changes in delta ferrite formations, grain boundary corrosion and sigma phase will arise. For avoiding this, pre- and postheat treatment are needed to prevent the metallurgical defects [3–7]. Moreover, joining of austenitic stainless steel under cryogenic or corrosive environment, the ferrite quantity to be minimized or controlled to avoid property degradation during service. It addition to this, it may also have a chance to sensitization in fusion welds [8–10].

4. Effect of friction welding in austenitic stainless steel

Due to high ductility and excellent corrosion resistance, austenitic stainless steel is increased in wide range of applications. Even though stainless steel is effectively used in commercial applications, problems have often been reported during welding operation. Many of the researchers are working.

It is observed that publications of most of the research papers are concerned with similar and dissimilar welding of 300 series grade of austenitic stainless steel. When comparing with fusion welding process, joining of austenitic stainless steel is increased subsequently using solid state process during the last decade. In metal joining process, wide categories of variables included in each circumstance and hence standardization of welding is difficult to find out in industrial aspect for avoiding such difficulties, research work is carried out to set the better performance in welding of austenitic stainless steel.

5. Research findings on friction welding of austenitic stainless steel to other material combinations

Many researchers have worked in friction welding which focused on joining similar and dissimilar combinations of austenitic stainless steel with different metal based alloys. Researchers have worked on dissimilar combinations of materials by resulting good bond strength under quality aspects in friction welding [11–14]. When a similar combination of austenitic stainless steel is performed, the value of tensile strength is decreased with increase in friction pressure [15]. Similarly, Paventhan et al. [16] studied a fatigue behavior by joining medium carbon steel and austenitic stainless steel by conducting experiments using bending fatigue testing. Further, experimental investigation was done on the friction welding of 6063 aluminum alloy with AISI 304 austenitic stainless steel by Sammaiah et al. [17] to determine the correlation between the microstructure and the joint strength. Similarly Fu et al. [18] investigated the welded joint of T2 copper and 1Cr18Ni9Ti stainless steel under the external electrostatic filed and the distributions of elements in weld zone (WZ) were analyzed in the welded joint. The influence of welding parameters on hot corrosion was examined by Arivazhagan et al. [19] to study the weldment and corrosion behaivor in elevated temperature on AISI 4140 and AISI 304. Subsequently, Sahin investigated a characterization of plastically deformed austenitic stainless steel by friction welding using statistical approach [20]. During fusion welding process of joining pure Ti to stainless steel, the formation of brittle intermetallic compounds developed in the weld metal. This problems lead to degrade the properties of weld joints. Muralimohan et al. [21] made an attempt to introduce thin Ni interlayer which overcomes the problems between Ti-SS by avoiding direct contact between two base metals. Satyanarayana et al. [22] studied the effect of austenitic-ferritic stainless steel combination in terms of microstructure and mechanical properties. The influence of strength and variations are compared together and its fracture behaviors are evaluated. Winiczenko and Kaczorowski [23] investigated the study of mechanical properties and microstructure of friction welded joint of ductile iron with stainless steel and studied the fracture morphology and phase transformations during friction welding. They also showed some enrichment of ductile iron with Cr and Ni atoms near to the weld joint through energy-dispersive X-ray spectrometry.

Some studies on similar and dissimilar combination of stainless steel materials are undergone to understand the impact and tensile strength behaviors using electron beam welding and friction welding [24]. The effect of tensile strength and impact at different loading rate was examined by Yokoyama et al. [25] on aluminum alloy and stainless steel using friction welding.

However, the combination of austenitic stainless steel to copper is very limited. This chapter shows a simple and novel approach to determine the welding parameters using Taguchi design by studying its mechanical and metallurgical properties.

6. Experimental details

The dissimilar joint combinations of austenitic stainless steel (304L) to copper material are taken into account for examination. The materials were chosen with a cylindrical rod of diameter 24 mm and length 75 mm. The surface is well polished and cleaned by using acetone. The chemical composition of base materials used for this experiment is shown in **Table 1**.

Element (%)	C	Si	Mn	P	S	Ni	Cr	Fe	Zn	Cu
AISI 304L	0.03	0.39	1.63	0.042	0.027	8.99	19.05	71.12	–	–
Copper	–	–	–	–	–	0.01		0.11	0.13	99.59

Table 1. Chemical composition of base materials.

The factors such as friction pressure, upset pressure, burn-off length and rotational speed is the main parameters involved in friction welding process. Taguchi's orthogonal array is a simple and largely useful method, for conducting experiments in a systematic way using a restricted number of experiments required for the investigation. The factors considered for the experimentation are listed in **Table 2**.

Factors	Levels		
	1	2	3
Friction pressure (MPa)	22	33	43
Upset pressure (MPa)	65	87	108
Burn-off length (mm)	1	2	3
Rotational speed (rpm)	500	1000	1500

Table 2. Experimental factors and their levels.

7. Surface appearance on weld

Figure 1 shows the appearance of 304L SS and copper combination made by friction welding. The welded joint between 304L and copper material reveals the formation of flash regions which contains predominantly copper, shown in **Figure 3**. This is has a result of the lower flow stress of copper, the heat generated during welding temperature makes softer in copper and starts flowing in terms of flash formation as compared to austenitic stainless steel side.

Figure 1. 304L SS-Cu welded joints.

Due to the ductility in copper material, flash is produced on the copper side with reduced length than stainless steel side.

8. Mechanical testing

8.1. Tensile testing

The mechanical test was carried out on the weld line by sectioning the welded samples. Tensile test was carried out at room temperature using a WAW1000E universal testing machine having the maximum load of 100 kN and 5 mm/min crosshead speed. The welded joints were machined for tensile testing according to ASTM E8 standard and joint strength was analyzed in the weld region. A scanning electron microscope (SEM) was used for observing the fractured surface on tensile tested sample as well as the type of fracture obtained for the material.

The input parameters developed based on Taguchi method were utilized to evaluate the friction welded joints by conducting experiments. The tensile strength results of the welded joints are listed in **Table 3**. Most of the samples are fractured in copper side and not in austenitic stainless steel side (**Figure 2**). This is due to high ductility in copper material that fracture results in copper material. Due to chemical in-homogeneity and microstructural changes, the tensile strength values, might have some variations with all the input parameters.

Among all the samples made by friction welding, the sample S21 and S7 are obtained as lowest (183 MPa) and highest (205 MPa) of tensile strength values respectively. Though the UTS of 304L and Cu base material has 647 and 232 MPa, the friction welded joint results with a maximum of 205 MPa. It indicates clearly that, the maximum tensile strength is more or less equal to the base material of copper. With higher the friction and rotational speed and low upset pressure, minimum tensile strength was observed. Similarly, higher tensile strength was obtained by increasing

S. No	FP (MPa)	UP (MPa)	BOL (mm)	Rotational speed (r/min)	UTS (MPa)	Fracture location
1	22	65	1	500	201	Copper
2	22	65	2	1000	196	Copper
3	22	65	3	1500	192	Copper
4	22	87	1	1000	203	Copper
5	22	87	2	1500	198	Copper
6	22	87	3	500	197	Copper
7	22	108	1	1500	205	Copper
8	22	108	2	500	204	Copper
9	22	108	3	1000	199	Copper
10	33	65	1	500	196	Copper
11	33	65	2	1000	192	Copper
12	33	65	3	1500	187	Copper
13	33	87	1	1000	198	Copper
14	33	87	2	1500	194	Weld
15	33	87	3	500	193	Copper
16	33	108	1	1500	200	Weld
17	33	108	2	500	199	Copper
18	33	108	3	1000	195	Weld
19	43	65	1	500	192	Copper
20	43	65	2	1000	187	Weld
21	43	65	3	1500	183	Weld
22	43	87	1	1000	194	Weld
23	43	87	2	1500	189	Copper
24	43	87	3	500	188	Weld
25	43	108	1	1500	196	Copper
26	43	108	2	500	195	Weld
27	43	108	3	1000	190	Copper

Table 3. Taguchi's L_{27} orthogonal array.

upset pressure and rotational speed of the welded joint. When the rotational speed increases, irregularities in faces are smoothed out by lowering frictional contact with increase in upset pressure. As a result, the material becomes plastic and achieved with complete contact with interface of the other material. Hence, any impurities present on interface remains trapped with increase in quality of the welded joint. Fracture analysis was done by using scanning electron microscopy (SEM) in tensile tested specimens are shown in **Figure 3**. It shows a dimple pattern in the whole width of the specimen and confirms the ductile mode of fracture.

http://dx.doi.org/10.5772/intechopen.70558

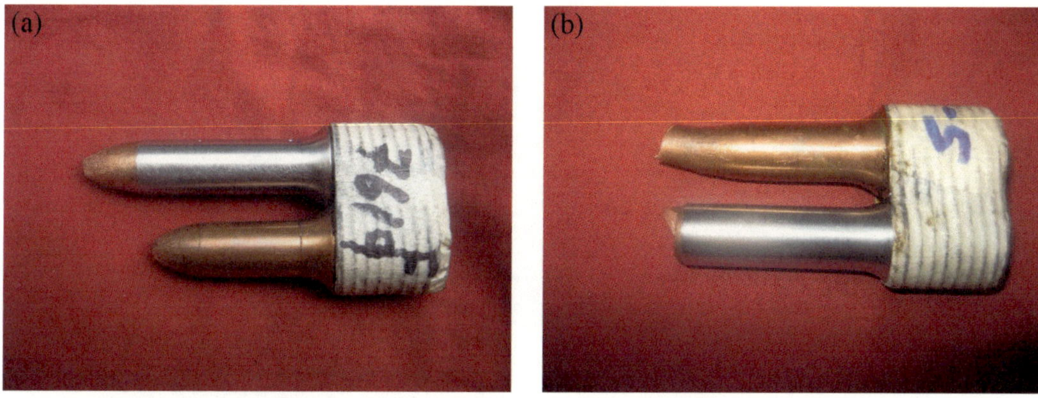

Figure 2. Tensile tested samples on 304L-Cu joint.

Energy dispersive analysis of X-rays (EDAX) analysis was carried out to study the phases that exist at the welding interface. The software permitted piloting the beam to scan along a surface or a line in order to achieve X-ray cartography or concentration profiles by elements [20]. SEM with EDAX analysis was carried out on the tensile fractured sample. SEM microstructure in the friction-welded 304L-Cu joint and EDAX analysis results are given in **Figure 4**, while the distributions of elements within the determined location are shown in **Table 4**. The analysis shows that the diffusion zones consisted of Cu and O atoms at the fractured surface. The diffusion zone is rich in Cu with a weight of 94.58%, followed by 5.42% O. Thus, the diffusion zone with a different element was confirmed with a copper material.

8.2. Impact testing

Impact testing was done by using Charpy V notch impact test machine to measure the impact toughness of joints at room temperature. The specimen size was 55 mm × 10 mm × 10 mm and the samples were prepared with ASTM standards. The samples to be tested were machined from the welded blocks. Notches were prepared precisely at the midpoint of the weld interface. The fractured surface of the impact tested sample was examined by using scanning elec-

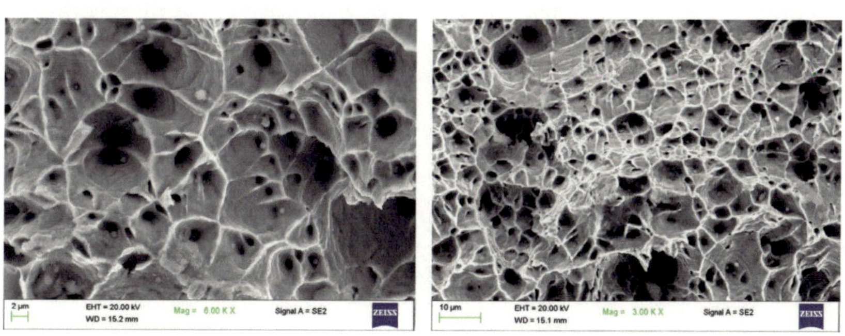

Figure 3. SEM image in tensile fractured sample on 304L-Cu.

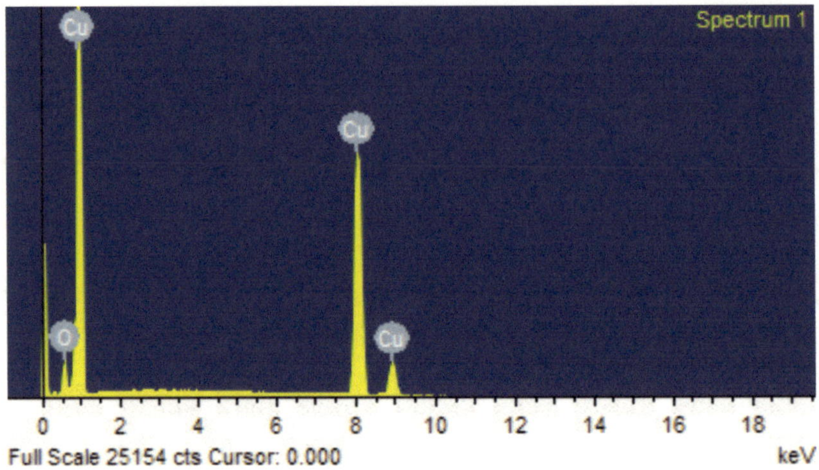

Figure 4. EDAX analysis on tensile sample in 304L-Cu.

tron microscope (SEM). Experiments are conducted using Taguchi's L_9 orthogonal array and the impact test results are presented in **Table 5**.

It was found that, the impact sample S7 has extremely low value of 4 J/cm^2 and the impact sample S2 has the highest value of 70 J/cm^2. The impact tested samples are shown in **Figure 5**. With high frictional pressure and a decrease in upset pressure, the impact toughness value is much reduced in weld interface. Due to the low upset pressure, the interface having irregularities and bonding toughness is much affected. At the same time, if increased with upset pressure and decreased with friction pressure, the value of toughness is drastically increased.

Due to increase in heat during friction, the presence of intermetallic layers are formed which results in poor weld strength. Based on the experiment, the energy absorbed by the material results with accumulation of copper particles on stainless steel side rather than with low energy absorbed by the material. The fractured surface of the impact tested sample was examined by using scanning electron microscope (SEM) to study the behavior of the material with different magnifications. The fracture shows ductile mode of fracture with coarse dimple features exhibited in the copper material by showing different magnifications shown in **Figure 6**. Toughness is higher with the low friction pressure as a resulted of higher deformation and failure occurs slightly away from the interface as evidenced by showing ductile failure. This supports the argument that the joint has good interface, which is formed with good toughness.

Element	Weight (%)	Atomic (%)
O K	5.42	18.54
Cu K	94.58	81.46
Total	100%	

Table 4. EDAX analysis in tensile fractured sample on 304L-Cu joint.

http://dx.doi.org/10.5772/intechopen.70558

Experimental run	Input parameters				Impact strength (J/cm²)
	Friction pressure (MPa)	Upset pressure (MPa)	Burn-off length (mm)	Rotational speed (r/min)	
1	22	65	1	500	16
2	22	87	2	1000	70
3	22	108	3	1500	66
4	33	65	2	1500	28
5	33	87	3	500	32
6	33	108	1	1000	8
7	43	65	3	1000	4
8	43	87	1	1500	30
9	43	108	2	500	38

Table 5. Impact test results on 304L-Cu joint.

SEM with EDAX analysis is shown in **Figure 7** and their elements observed are listed in **Table 6**. The diffusion zone observed on the impact tested samples and shows with rich Cu in the fractured sample which is occurred in copper material rather than the stainless steel material.

Figure 5. Impact tested samples on 304L-Cu joint.

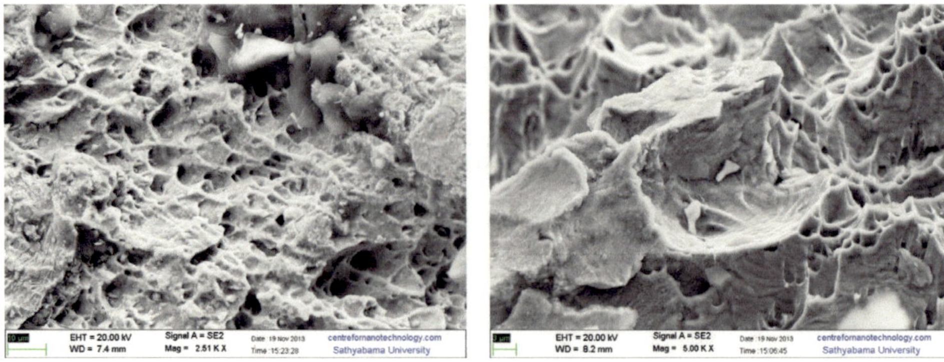

Figure 6. SEM image in impact fractured sample on 304L-Cu joint.

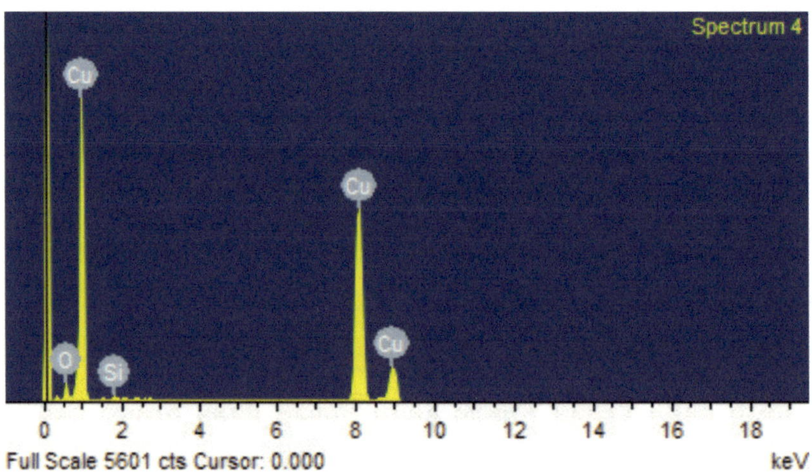

Figure 7. EDAX analysis on impact sample in 304L-Cu.

Element	Weight (%)	Atomic (%)
O K	3.61	12.90
Si K	0.32	0.65
Cu K	96.07	86.45
Total	100%	

Table 6. EDAX analysis in impact fractured sample on 304L-Cu joint.

8.3. Hardness testing

Vickers micro-hardness measurements were made across the weld on all samples to identify the strength in the three microstructural zones such as weld zone (WZ), base metal zone (BMZ) and heat affected zone (HAZ) in the respective materials. Vickers micro-hardness test was carried out across the weld interface using a load of 500 g and dwell time of 15 s along the weld interface. Vickers micro-hardness measurements were carried out accordance with ASTM E384-09 and ASTM E407-99 standards, respectively. The hardness values are taken at each location while the average of three readings was taken for analysis.

In the case of hardness in welded joint of austenitic stainless steel and copper, it was not possible to take the hardness in the weld zone, as the weld is just a sticky mushy zone. Hardness variation was obtained using 500 g load by Vickers micro-hardness testing and measuring locations with 0.5 mm intervals taken into consideration. The hardness variations in horizontal distance to the center in the welding interface of the joints are shown in **Figure 8**.

It could be observed that, the hardness values of copper near to the weld interface are slightly increased when compared to the base material of copper. At the same time, hardness value of 304L SS slightly decreased near to weld interface when it compared to the base material of stainless steel. Due to heat dissipation at weld interface, intermetallic layers and thermal diffusivity occur which causes hardness variations.

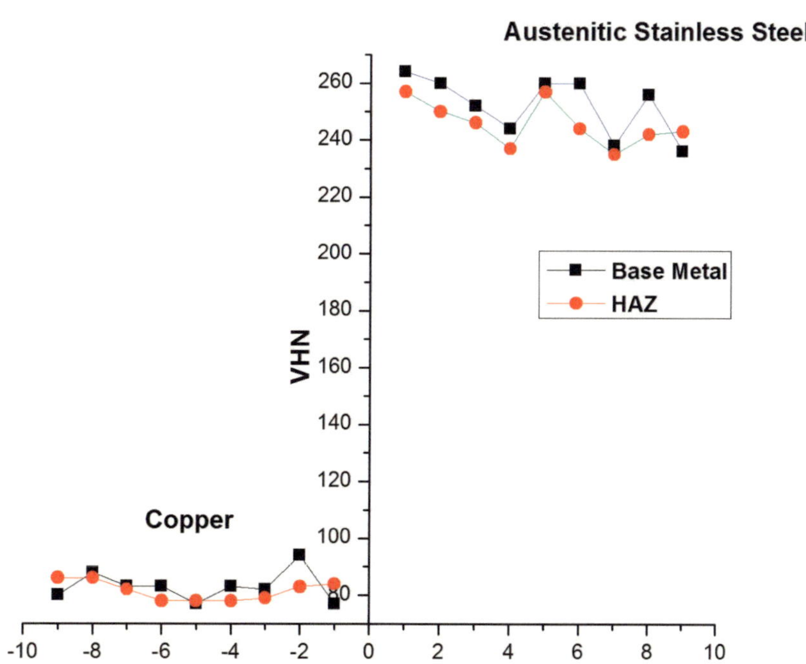

Figure 8. Hardness graph in 304L-Cu joint.

9. Metallographic examination

9.1. Optical microscopy

The optical microscope study was carried out to examine the grain behaviors in the interfaces and heat-affected regions. The microstructures were examined by sectioning the weld samples, parallel to the radial direction and the specimens were prepared according to standard metallographic procedures. The welded surface of the samples was ground with 1200 grinding paper and polished with 1 μm diamond paste and the samples were etched with a vilella's regent (5 ml HCl, 1 g picric acid, 100 ml ethanol and 2 drops zephiran). The specimen is well-polished and etched by 10% oxalic acid. The welded joint was examined using a metallurgical microscope and microstructural behaviors were analyzed in base metal, heat affected zone (HAZ) and the weld zone.

Figure 9 shows the micrograph showing microstructures in weld region as well as heat-affected zone and parent metal across the interface. Due to the heat applied during the welding operations, the flash thickness was varied from one another resulting in plastic deformation at the interface. Due to cylindrical rods with a circular geometry, the rotational speed affects the frictional pressure from the weld center to the surface of the sample in radial directions. The parent metal of copper was observed with coarse alpha grains and by heat-affected zone, and the grains were recrystallized due to heat generated in the interface of weld region. In austenitic stainless steel, the parent metal was observed with carbide particles and annealed twin boundaries, whereas in the heat affected zone, recrystallized grains were appeared.

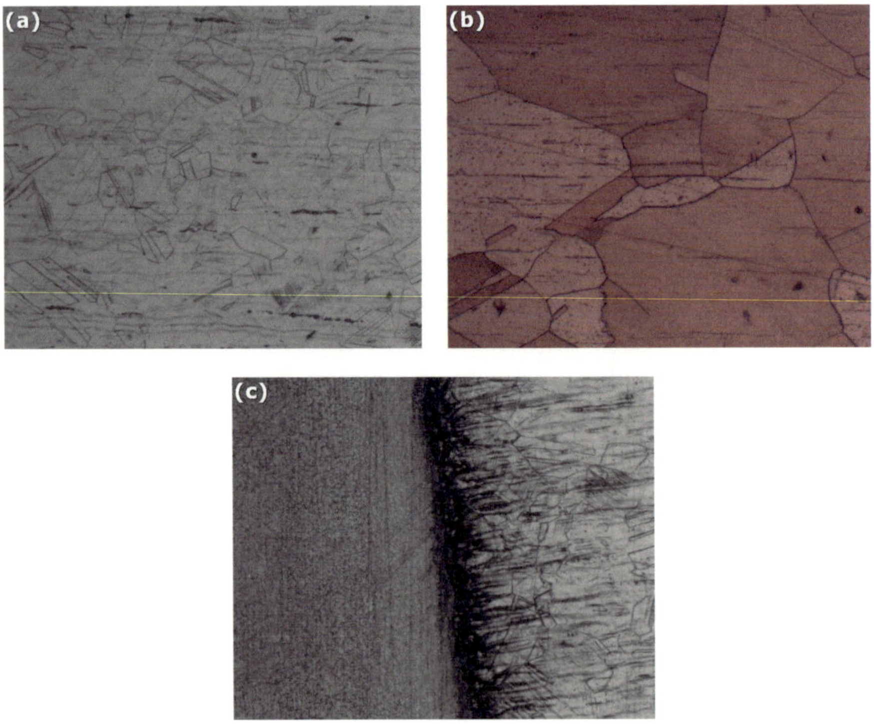

Figure 9. Microstructural observation of welded samples (a) Austenitic Stainless Steel, (b) Copper and (C) Weld Interface.

http://dx.doi.org/10.5772/intechopen.70558

9.2. Atomic force microscopy

Atomic force microscopy (AFM) is a powerful technique which can allow direct spatial mapping of surface morphology having nanometer resolution. The roughness images were taken over by the integrated optical microscope and operated in tapping mode using silicon probes. Topographic and phase images were achieved concurrently by a resonance frequency of approximately 300 kHz for the probe oscillation and a free-oscillation amplitude of 62 ± 2 nm. The microstructure of interphase layer of dissimilar material is seen in atomic force microscopy. The maximum roughness shows with 45 nm in 304L, 236 nm in copper and 246 nm on the interface. The maximum roughness in the interface zone has more or less equal to the same roughness as of copper.

From the roughness graph and 3D images (**Figure 10**), was observed that the difference between the average roughness of dissimilar material is very less and negligible in the interface region. When studying roughness size, the parent materials of 304L SS and copper are having peaks in the range of 15–35 and 30–90 nm, respectively. In the welding zone, the peak appeared is with a range of 60–130 nm showing considerable increase in roughness.

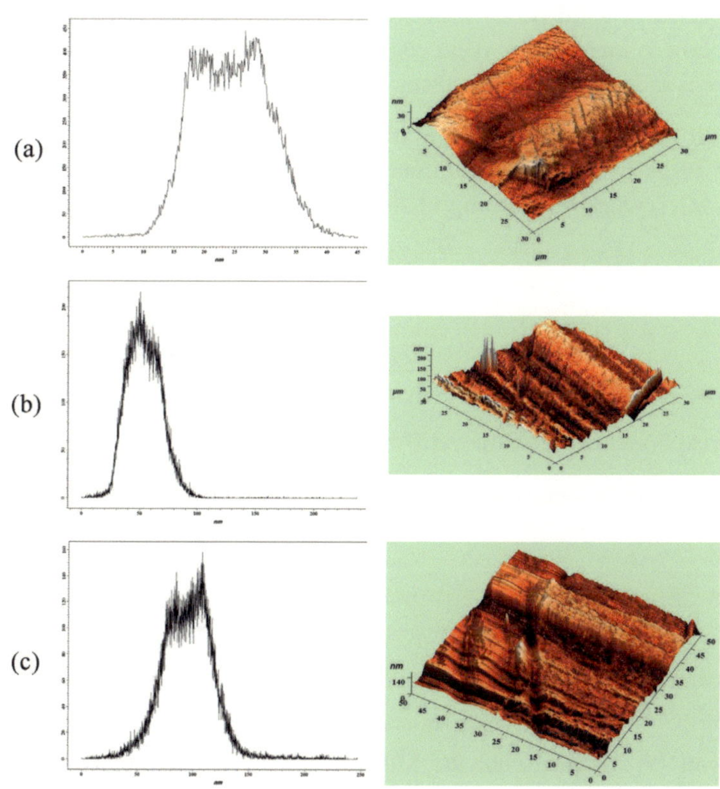

Figure 10. AFM histogram and 3D image in 304L-Cu (a) Austenitic Stainless Steel, (b) Copper and (c) Weld Interface.

10. Conclusion

In this study, the welding characteristics under different welding parameters are taken into account. From the mechanical and metallurgical characteristics on friction welding of austenitic stainless steel and copper, the following conclusions can be drawn.

- During friction welding, the metal tends to decrease in length of copper by showing a flash formation rather than stainless steel side.
- Tensile strength was achieved with a maximum of 205 MPa. The bond strength is achieved nearing to base material of copper material with increase in upset pressure.
- Energy absorbed by the welded sample is varied from 4 to 70 J/cm^2. The bond strength is mainly depends upon one of the important welding parameter called upset pressure. This kind of statement is well suites in current study of impact. When the upset pressure is lowered, the impact toughness is much decreased to 4 J/cm^2. But in case of high upset pressure, the resulting value is drastically increased to 70 J/cm^2.
- Micro-hardness measurements in 304L-Cu results with absence of hardness value in weld zone. Due to sticky layer at the interface of stainless steel as well as copper joint, the weld zone is negligible and measured values in the HAZ and their base materials, respectively.
- Due to the absence of weld zone in 304L-Cu, surface roughness was studied using atomic force microscopy to identify the weld region. Measurement of roughness values made in weld interface is more or less equal to copper material.

Though austenitic stainless steels are used more than any other grades, the joining efficiency is differed with respect to welding parameters. Finally, it is suggested that there is a large scope in the research work on welding of 300 series grade with wide range of applications.

Author details

Shanjeevi Chinnakannan

Address all correspondence to: shanjeevii@gmail.com

Department of Mechanical Engineering, Veltech Dr. RR & Dr. SR University, Chennai, India

References

[1] Anik S, Tulbentci K. Istanbul: Handbook of Welding Technique, Gedik Welding Publications, Istanbul, Turkey.

[2] Satyanarayan VV, Madhusudhan RG, Mohandas T. Dissimilar metal friction welding of austenitic-ferritic stainless steels. Journal of Materials Processing Technology. 2005;**160**:128-137

[3] Tylecote RY. The Solid phase welding of metals, London: Edward Arnold Ltd; 1968. pp. 1-150.

[4] Tsuchiya K, Kawamura H. Mechanical properties of CuCrZr alloy and SS316 joints fabricated by friction welding method. Journal of Nuclear Materials. 1996;**913-7**:233-237

[5] Sahin AZ, Yibas BS, Ahmed M, Nickel J. Analysis of the friction welding process in relation to the welding of copper and steel bars. Journal of Materials Processing Technology. 1998;**82**:127-136

[6] Yılmaz M, Col M, Acet M. Interface properties of aluminium/steel friction welded components. Materials Characterization. 2002;**49**:421-429

[7] Baeslack WA, Puquette DJ, Savage WF. Effect of ferrite on stress corrosion cracking in duplex stainless steel weld metals at room temperature. Corrosion. 1979;34-46

[8] Szumachowki ER, Reid HF. Cryogenic toughness of SMA austenitic stainless steel weld metals part-1 role of ferrite. Welding Journal. 1978;**57**:325s-333s

[9] Satyanarayana VV, Madhusudhan Reddy G, Mohandas T, Venkata Rao G. Continuous drive friction welding studies on AISI 430 ferritic stainless steel welds. Science and Technology of Welding and Joining. 2003;**8**(3):184-193

[10] Satyanarayana VV, Madhusudhan Reddy G, Mohandas T. Continuous drive friction welding studies on AISI 304 austenitic stainless steel welds. Materials and Manufacturing Processes. 2004;**19**(3):487-505

[11] Ananthapadmanaban D, Seshagiri Rao V, Abraham N, Prasad Rao K. A study of mechanical properties of friction welded mild steel to stainless steel joints. Materials and Design. 2009;**30**:2642-2646

[12] Hascalik A, Unal E, Ozdemir N. Fatigue behavior of AISI 304 steel to AISI 4340 steel welded by friction welding. Journal of Materials Science. 2006;**41**:3233-3239

[13] Subhash Chander G, Madhusudhan Reddy G, Venugopal Rao A. Influence of rotational speed on microstructure and mechanical properties of dissimilar metal AISI 304-AISI 4140 continuous drive friction welds. Journal of Iron and Steel Research. 2012;**19**(10):64-73

[14] Dey HC, Ashfaq M, Bhaduri AK, Prasad RK. Joining of titanium to 304L stainless steel by friction welding. Journal of Materials Processing Technology. 2009;**209**:5862-5870

[15] Sathiya P, Aravindan S, Noorul Haq A. Mechanical and metallurgical properties of friction welded AISI 304 austenitic stainless steel. International Journal of Advanced Manufacturing Technology. 2005;**26**:505-511. DOI: 10.1007/s00170-004-2018-6

[16] Paventhan R, Lakshminarayanan PR, Balasubramanian V. Fatigue behavior of friction welded medium carbon steel and austenitic stainless steel dissimilar joints. Materials and Design. 2011;**32**:1888-1894

[17] Sammaiah P, Suresh A, Tagore GRN. Mechanical properties of friction welded 6063 aluminum alloy and austenitic stainless steel. Journal of Materials Science. 2010;**45**:5512-5552. DOI: 10.1007/s10853-010-4609-y

[18] Fu L, Du SG. Effects of external electric field on microstructure and property of friction welded joint between copper and stainless steel. Journal of Materials Science. 2006;**41**: 4137-4142

[19] Arivazhagan N, Singh S, Prakash S, Madhusudhan Reddy G. An assessment of hardness, impact strength, and hot corrosion behavior of friction-welded dissimilar weldments between AISI 4140 and AISI 304. International Journal of Advanced Manufacturing Technology. 2008;**39**:679-689

[20] Sahin M. Characterization of properties in plastically deformed austenitic-stainless steels joined by friction welding. Materials and Design. 2009;**30**:135-144

[21] Muralimohan CH, Ashfaq M, Ashiri R, Muthupandi V, Sivaprasad K. Analysis and characterization of the role of Ni interlayer in the friction welding of titanium and 304 austenitic stainless steel. Metallurgical and Materials Transactions A. 2016;**47A**:347

[22] Satyanarayana VV, Madhusudhan Reddy G, Mohandas T. Dissimilar metal friction welding of austenitic-ferritic stainless steels. Journal of Materials Processing Technology. 2005;**160**:128-137

[23] Winiczenko R, Kaczorowski M. Friction welding of ductile iron with stainless steel. Journal of Materials Processing Technology. 2013;**213**:453-462

[24] Madhusudan Reddy G, Srinivasa Rao K. Microstructure and mechanical properties of similar and dissimilar stainless steel electron beam and friction welds. International Journal of Advanced Manufacturing Technology. 2009;**45**:875-888. DOI: 10.1007/s00170-009-2019-6

[25] Yokoyama T, Ogawa K. Impact tensile properties of 6061 aluminium alloy to SUS 304 stainless steel friction-welded butt joints. Welding International. 2003;**17**(7):514-523

Chapter 10

Study of Fracture Mechanisms at Cyclic Fatigue of Austenitic Steels Used in Nuclear Reactors

Tamaz Eterashvili

Additional information is available at the end of the chapter

http://dx.doi.org/10.5772/intechopen.70953

Abstract

The presented work deals with the possible reasons of nucleation and propagation of macro-, meso-, and microcracks after low-cycle fatigue (LCF) tests of stainless, austenitic steels at room temperature. This research will also support the solution of some important problems of steel modeling and their application to EURATOM, FUSION, TACIS, PHARE, CORDIS, PERFECT, AMES, FP 7, and GEN IV programs. The scanning electron microscopic (SEM), X-ray, and transmission electron microscopic (TEM) examinations and statistical analysis of the samples which undergone the low-cycle fatigue (LCF) showed that the average length of slip bands and micro-components of macrocrack are equal, and they are always parallel to each other, indicating their crystallographic character. The microcracks in these samples, caused by the residual stresses after the LCF, were studied in the TEM samples after their preparation to reveal the features of microcrack initiation and nucleation. It was shown that microcracks propagate along slip bands and change their propagation direction at the boundaries of grains and subgrains. This confirms that microcracks, as well as micro-components of mesocrack, are crystallographic. Mathematical calculations of elastic-plastic model are also performed which showed that the length of plastic zone, according to our assessments, is equal to 110–120 ***d***, where ***d*** is a width of crack opening.

Keywords: fracture, fatigue, SEM, TEM, slip band, twin, microcrack, microstructure, plastic zone

1. Introduction

Among others, the problem of low-cycle fatigue (LCF) fracture is of particular interest, and a number of papers and monographs are dealing with this problem. It is known that nucleation of microcracks in the austenitic, fine-grained steels was observed in slip bands. In the recent publications, the key role in the destruction of the material is assigned to precipitates and grain boundaries as well, such as MC-type carbides that precipitate mainly on grain boundaries. It was also shown that nucleation of fatigue cracks occurs at nonmetallic inclusions; however, the trajectory of further propagation of the crack passes through the sites of local stresses with equal magnitudes. The authors of [1] assumed that in stainless steel 304, at different fatigue cycles, size of nucleated crack depends on the intensity of stress. Stress intensity factor (K), to influences the plastic zone size as well, resulting in a complicated dependence of the latter on K [2]. In work [2], it carried out a modeling of plastic zone behavior during a ductile-brittle fatigue and found a certain correlation between the plastic zone size ahead of the arrested crack and the stress intensity factor. Local and global anisotropy considerably affects mechanical behavior of materials, mostly when the dimensions of microcracks and grains are comparable. There are also indications that the cracks of the above dimensions propagate with considerable deviations and more intensively at the initial stage of crack growth. Notwithstanding, this "anomalous behavior" is usually attributed to the crystallography of grains; its reasons are not yet fully understood. The varying orientations of grains may create nonuniform stress field. Generally, the influence of the fields is not taken into account for characterization of fracture process. Usually, the growth of small cracks is described in terms of linear elastic fracture mechanics (LEFM); the considerations of the latter are applicable for large cracks as well. Therefore, the local anisotropy of grain is neglected in Ref. [3], and the influence of local alteration of elasticity modulus on mechanical fracture of "microstructurally small" cracks has been estimated, in particular, of the stress intensity factor (SIF) K and its variation with regard to random crystallographic orientations of grains. The microcrack propagation process and its correlation with the microstructure of the steel are also quite disputable. Despite numerous experimental results on alteration of microcrack propagation direction, there is no convincing hypothesis explaining the real reasons for the process. The main goal of the proposed work is the careful study of the problems described above. For this purpose different grades of chromium austenitic steels have been examined which are the prospective materials for structural parts subjected to extreme cyclic loading and severe environmental conditions.

2. Experimental

2.1. Materials and their treatment

Standard cylindrical samples for cyclic tests were cut off from the thermally treated rods using lath and milling tool. Some of the samples were of square cross section. Working parts of the samples were polished for metallographic analysis. Austenitic grains were revealed by chemical etching. After completion of 20, 30, 40, and 50% of cyclic deformation of fatigue tests, at different amplitudes of deformation, ±0.05 and ±0.5, the evolution of slip lines was studied (**Table 1**). The characterizations were carried out by scanning electron microscope. Fractographic characterizations were carried out after rupture of the samples using a scanning microscope. The local plastic deformation rate and the magnitude of the respective stresses within the micro-area of a grain were determined. The samples for transmission electron microscopic (TEM) investigations were prepared by mechanical polishing and then electrolytic and/or ion bombardment

	Content wt%									
Steel grades	**C**	**Si**	**Mn**	**Cr**	**Ni**	**Ti**	**Mo**	**Nb**	**S**	**P**
X18H10T-AISI-321	0.08–0.12	0.7–0.8	1–2	17–19	9–11	0.5–0.7	—	—	0.015–0.02	0.03–0.035
X18H12Б-AISI 347	0.06–0.08	0.8–1.0	1–2	17–19	11–13	—	—	1.0–1.1	0.2–0.025	0.03–0.035

Table 1. Chemical composition (wt%) of the investigated steels.

techniques. In order to reveal the location of stress concentrations, electron beam defocusing was applied. Via stereographic and crystallographic analysis and computer simulation of the respective schemes, the following were determined: the direction of the slip lines, modes of deformation twins, stacking faults, and microcracks. On the basis of the obtained experimental results, some possible theoretical models of microcrack initiation mechanisms have been considered. The above heat treatments lead to the formation of big austenitic grains that supported the study of plastic processes taking place within one austenitic grain.

2.2. Low-cycle fatigue (LCF) tests

LCF tests were performed at room temperature. Two regimes were used: (1) The same amplitude of deformation ±0.05 mm and different frequencies of cycling, 0.33 and 0.45 Hz, and (2) frequency, 0.45 Hz, and amplitudes ±0.05 and ±0.1 mm. The cylindrical samples were loaded cyclically with the magnitude of loading 100–3000 kg. For scanning electron microscopic (SEM) study, in some cases, the V-notched plate-like samples were also used. All samples were subjected to continuous cyclic loading.

3. Results

3.1. Scanning electron microscopic (SEM) investigations

The SEM study of the samples revealed a peculiar polycrystalline structure of austenite. The sizes of austenite grains vary from 30 to 70 μm, depending on grade and treatment of the steel. In some large austenite grains, the recrystallization twins and variety of shapes of grains were also observed. The grains turn rounded; acute angles and joints between the grains become smoother. The majority of the metallographic sections showed the austenite grains with the diameters of 30–70 μm. After cycling (n = 2000), a macrocrack was formed on the metallographic section, and a 150–250 μm width zone with the plastic features was also detected at the tip and the both sides of the macrocrack. After subsequent fatigue tests, the width of the zone increased up to 250–300 μm. Within the plastic zone, in addition to the heavily deformed grains, some partially deformed or even non-deformed grains were revealed. As a rule, they were formed at the boundaries of the large grains and twins and at joints of grains. In addition to the localized slip bands in the plastic zone of some grains, an intensive plastic flow was detected in the form of parallel slip bands covering the part of the grain.

Further cyclic fatigue deformation, in addition to the existing slip bands, caused the creation of some new ones (**Figures 1** and **2**). The variation of lengths of the slip bands and the respective statistical curves (**Figure 3**(n_1)) revealed that most of the bands are of 10–30 μm length. Further tests lead to the formation of the additional slip bands that, in contrast to the previous ones, are shorter and less neat, while their maximal number remains the same (**Figure 3**(n_2)). Further fatigue tests caused a minor (20%) increase in the lengths of the slip bands. The widths of the bands increased 2–3 times. However, in spite of the latter, the slip bands do not cover the whole width of the austenite grain. The measurements of angles between the axis of loading and the slip bands showed that the first slip bands are oriented at an angle of 30–70°. However, the majority of slip bands were formed along the direction of maximal tangent stresses (45°). The direction of the slip bands, formed at later stages of the test, nearly or completely coincides with the direction of the macrocrack propagation.

3.2. Macrocracks

As it was noted above, at the crack tip and around it, the slip bands and isolated mesocracks were observed. Most frequently, the isolated mesocracks were formed near the slip bands. However, in some cases, they were observed inside small grains as well. It should be noted particularly that deformation in plastic zone flows rather heterogeneously. In the places of deviation, a plastic deformation takes place in the form of slip bands (**Figures 2** and **4**). The macrocrack consists of individual micro-components.

In the 30–50 μm area near the crack tip, a plastic deformation of a higher rate than at both sides of the crack was detected. The latter is exhibited in the form of the enhanced homogeneous slip in the surface plain of the sample and in the direction perpendicular to it, and the localized slip bands become coarser. The variation of the lengths of micro-components of the macrocrack showed that after n_2 = 4000 cycles their lengths vary between 10 and 30 μm. However,

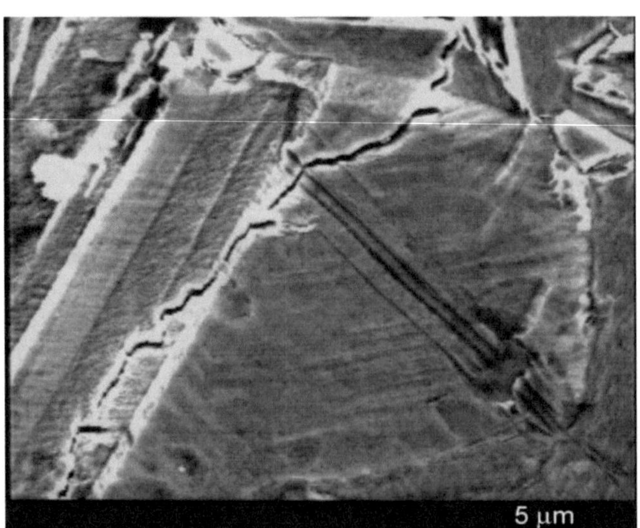

Figure 1. Localized slip bands near the large carbides and other inclusions (X18H10T mode II).

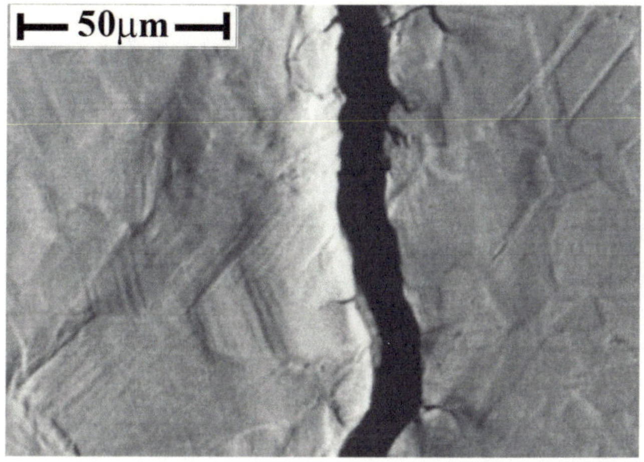

Figure 2. Intensive plastic flow, near a macrocrack in the steel X18H12Б (mode I). (SEM image).

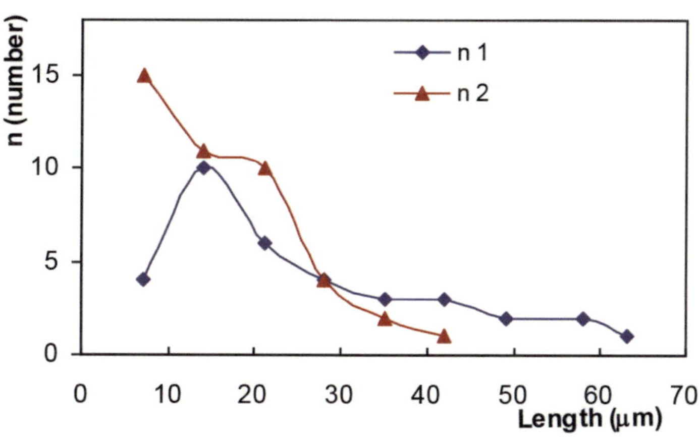

Figure 3. Statistical distribution of the lengths of slip bands (n_1 = 2000 and n_2 = 4000 (X18H12Б)).

the lengths of the micro-components decrease after subsequent tests. The angles of deviation of the micro-components from the axis of cycling were also measured. As it is evident, the majority of micro-components deviate at an angle of 80–90°. The macrocrack passes through the twin/matrix boundaries without any significant deviations as well. It was also revealed that the dimensions of micro-components of a macrocrack and microstructure elements of the steel better coincide in the lower range of lengths. Thus, in the growth and propagation, the macrocrack correlates with the microstructure elements of the steel. Thus, one may assume that micro-components of macrocrack are influenced by the crystallography of the material.

3.3. Isolated mesocracks

In plastic zone, some individual, isolated mesocracks were also detected in addition to the slip bands (**Figure 4**). They mainly nucleate at (or on) the grain boundaries with no apparent signs

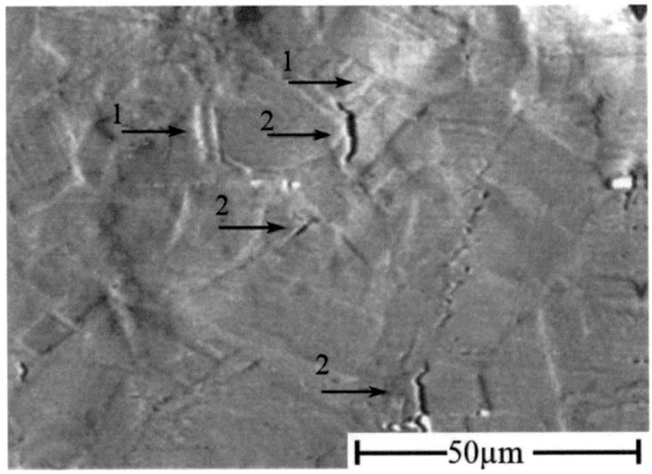

Figure 4. Plastic zone ahead of the fatigue macrocrack ((1) slip bands, (2) microcracks).

of plastic deformation (e.g., slip bands), while the isolated mesocracks were quite frequently observed within the small grains as well. In some areas, they are grouped, while in the others, the distance between them is 50–60 μm. As the isolated mesocracks are located in the plastic zone, the distance between the macrocrack and the utmost mesocrack is controlled by the dimensions of a plastic zone (250–350 μm). Here, we also have to note the experimentally observed fact that the directions of the isolated mesocracks and the slip bands often (but not always) coincide. In the grains in which the isolated mesocracks were formed, no localized slip bands were observed. Thus, it should be supposed that the mesocracks were formed near the border of plastic zone. In the majority of cases, the orientation of mesocrack coincides or is close to the main direction of macrocrack [4]. Analysis of different areas of the zone and measurements showed that the size of the plastic zone is related to that of macrocrack, and an increase in macrocrack size increases plastic zone size. The maximal size of plastic zone ahead of macrocrack tip detected in our experiments was equal to 250–350 μm (**Figure 4**).

The macrocrack is localized in a narrow zone 600 μm wide in which isolated meso- and micro-cracks and slip bands are observed. Clearly, the size varies depending on chemical composition and heat treatment of the steel; however, the correlation between macrocrack size and the plastic zone size still exists. The shape of the plastic zone is close to a spherical or compressed ellipse. Theoretical solution of this problem is foreseen in the proposed project based on an elastic-plastic model using the linear theory of elasticity.

3.4. The elastic-plastic model

The simple assumption on fatigue failure mechanism in the conditions of plain strain leads to the formation of a small plastically deformed volume ahead of microcrack, called the crack tip plastic zone. The zone is cylindrical with the radius R and the axis is oriented as shown in **Figure 5**. If the radius (R) of the relaxation zone is less than the crack length (L), the stress in the middle of the crack practically will not be affected by the plastic relaxation.

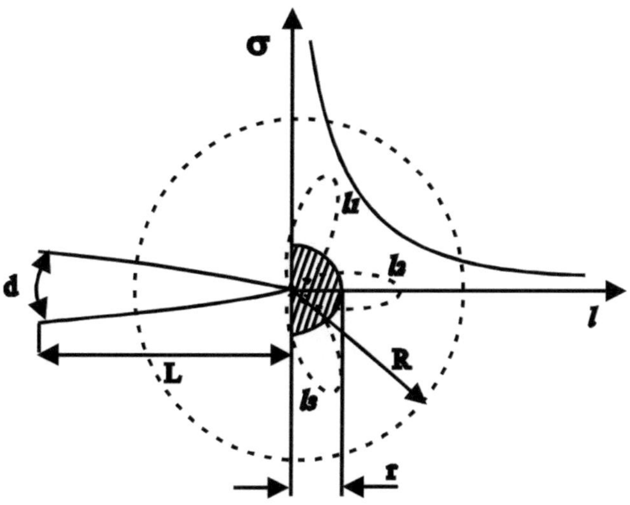

Figure 5. Theoretical scheme, of the cylindrical plastic zone, ahead of the crack tip. Here, *r* s a radius of destruction zone, and l_1, l_2, and l_3 are dimensions of different plastic characteristics (slip bands, twins, and microcracks).

In addition, the crack width (*d*) remains the same. Our experiments show that the edges of the cracks near the crack tip are parallel to each other. According to the linear elastic fracture mechanics, the magnitude of the stress applied perpendicularly to the microcrack growth direction in the vicinity of a crack tip is of the form

$$\sigma_F^2 = K^2/2\pi \mathrm{R} \quad (1)$$

In order to evaluate the size of the plastic zone, let us suppose that on the interface between the plastic and elastic zones the magnitude of stress (σ_E) is equal to the yield strength of the material (σ_F). However, based on the above supposition, we are considering only the interface between plastic and elastic zones. Here, it should be noted that the formula contains a stress intensity factor (*K*), being quite difficult to determine

$$R = \frac{K^2}{2\,\pi\sigma_F^2} \quad (2)$$

unambiguously (in addition to the stressed state, the factor *K* depends on the shape of the microcrack as well). In the present work, an attempt is made to exclude this factor from Eq. (2). To characterize the material resistance to crack propagation, a critical value of crack opening (*d*) may be used instead of *K*. According to the reported results of a number of authors, this value may be used in the cases of substantial plastic deformations as well. When the material with several slip systems (in our case of FCC crystal it has) is analyzed, the elastic stresses occurred near the crack tip are almost completely relaxed in the plastic zone, while in the elastic zone, they still exist, and *K* characterizes the stressed state. As a result, the crack tip turns blunt, and the edges go parallel to each other, except the area at the very tip. In plastic zone, the dislocations are in equilibrium state because of yield stress (σ_F) and the external stress (σ) applied from the side of the microcrack:

$$I = \text{nb}\sigma_F \tag{3}$$

Here, I is a Rice integral [5] which determines the energy flow to the crack tip or the body potential energy change after the crack length is varied. Especially, I determines a free energy per unit length of the crack ($I = -\partial w/\partial h$). In the cases when $R \ll L$, that is, for small plastic zones, we arrive at the following expression for the planar stressed state:

$$I = K^2/E \tag{4}$$

Here, E is a Young modulus. However, as it follows from Rice integral, I determines the energy dissipated in the process of crack propagation and is related to the crack opening (d):

$$d = I/\sigma_F \tag{5}$$

$$I = \text{d}\sigma_F \tag{6}$$

The formulae (4) and (6) lead to the exclusion of I, and we arrive at the following formula for K:

$$K^2 = \text{dE}\sigma_F \tag{7}$$

From Eqs. (7) and (2), we will also have for R

$$R = \frac{\text{dE}}{2\pi\sigma_F} \tag{8}$$

Thus, the plastic zone size can be determined this way, knowing E and σ_F of the material. Calculations of R for the steel with $E = 195{,}000$ MPa and $\sigma_F = 280$ MPa showed that for this steel $R \cong 110\,d$, that is, it is equal to 300–350 μm that coincides with our experimental measurements ($r \cong 0.1\,R$) [6].

4. TEM examination structure changes after LCF

As it was shown above, the size of the plastic zone ahead of macrocrack amounts to 250–300 μ (spherical or ellipsoid). For TEM studies we used samples with the diameter of 3 mm and initial thickness 200–250 μ, which subsequently were mechanically ground to 200–250 μ and then electrochemically polished to 500–600 Å. Comparison of this dimension with those of the plastic zones makes it clear that we are not able to study the processes taking place ahead of macrocrack. This means that all the processes in plastic zone can be detected and studied except formation of meso- and microcracks, due to the dissolution of areas adjacent to micro-crack during the electrochemical polishing. On the other hand, so big tension is concentrated just ahead of microcrack that during thinning the sample is destructed. However, internal

http://dx.doi.org/10.5772/intechopen.70953

tensions still remain in some areas of the prepared thin film, and after relaxation, they can be revealed in the places of peak tension. For example, in such places it may form structure defects including nanocracks and microcracks. To understand the latter, one should take into consideration that all kinds of fracture, brittle, ductile, and fatigue, flow via dislocation movement that provides formation of slip bands, stacking faults, and deformation twins. Thus, the reason for the formation of the above defects is microflow. In the cases when the microflow is impossible because of sufficient tension, and when the possibility of formation of the above defects is exhausted, a microcrack is nucleated. Subsequently, peak tensions relax in the places of microcrack formation. That is why TEM samples were held for 0.5–1.0 hours at room temperature for redistribution of internal tensions and formation of microcracks and only after that they were examined in the TEM. The stacking faults, traces of dislocation movement, and sometimes even slip bands are also present in addition to twins. The trace and stereographic analysis showed that in all the investigated cases the slip is realized in {111} planes along <110> directions. The microstructure of the steel is substantially changed, and the characteristic polygonal structure of austenite occurs. In the beginning the polygon boundaries are thin and then gradually thicken when deformation increases. The sizes of the polygons vary from 1 to 20 μm. The average orientations of their elongations coincide with the slip direction (<110>). However, disorientation between the polygons is easily observed which ranges from 2 to 4°. The same is with azimuthal disorientation during cyclic deformation. The polygons also differ from each other by the dislocation density and are detected only at polygon boundaries. Further study of fatigue plastic deformation of the investigated steels showed that increase in a number of cycles causes intensive structural transformations in them. The trace and diffraction analysis showed that the slip takes place along the common planes and directions for austenite ({111} and <110>) (**Figure 7**). In addition to the slip bands in some particular grains, a formation of cellular dislocation structure begins (**Figure 8**). At these stages, a dimension of some carbide precipitates decrease. Here should be noted that according to the internal structure of the steel (**Figures 7–9** and **11–13**), in addition to the regular spots, diffuse intensities occur as well (**Figure 9b**). The firstly formed dislocation cells are larger; their dimensions vary from 0.02 to 0.25 μm. The dimensions of dislocation cells gradually decrease after consequent deformations reaching 0.01 μm and, consequently, transform into dislocation balls (**Figure 8**). In some areas, dislocation bundles are formed in which particular dislocations cannot be distinguished. On the one hand, it facilitates the strengthening of the material, and on the other hand, it indicates at enabling other systems of stress relaxation. Along the formation of cells, the process of deformation twinning is going on. Thus, while in one grain, bundles are observed, and in others deformation twins may be observed. Each microstructure element deforms according to its crystallographic orientation to the applied stresses. In the twins as well as in the areas between them, a sufficiently high density of dislocation remains just after the area is completely covered by twins which are almost uniformly distributed within the grain. However, later, an increase of deformation rate increases the density of twins. Further increase of deformation rate provides formation of additional twins angularly with respect to the initial ones, that is, other twin modes come to action. Therefore, in some grains only one mode is detected, while in others two or three different twinning modes are operative. Obviously, just this twinning mode is most probable, and one may suppose that another twinning system comes to action after strengthening of an active one. Thus, twinning provides relaxation of tensions. After the formation of twins in the majority of grains is over, their size reduction takes place, and thereby the process of fragmentation of the structural elements of the steel starts. Along the directions <110> and <112> (directions of slip and twinning), azimuthal disorientation approaches 20^0 and more. This makes diffraction pattern complicated, and its indexing is ambiguous. In addition, on the diffraction patterns, the extra reflections caused by twins and carbide precipitates appear. At this stage nucleation of martensitic phase takes place, and the respective diffraction spots emerge (**Figure 10**). As the martensitic phase nucleates in deformed austenite, the shape of martensite crystals are the same as of deformation twins. Moreover, they are aligned along the twin propagation direction. To reveal them a the dark-field technique was used which showed that the lengths of martensite crystals are comparable with those of twins. Their lengths are limited by deformation twins and other martensite crystals. The latter becomes apparent through further fining of martensite and deformation twin crystals. The martensite crystals in metastable austenitic phase formed during deformation. The habit pole of the martensitic crystals is situated near the pole of deformation martensite.

5. Characterization of destruction zone ahead of microcrack

5.1. Interaction of microcracks with subgrains

As it was mentioned above, the TEM studies after low-cycle fatigue tests in the TEM samples showed that microcracks are formed in the places of stress concentration and the nucleated microcracks are of different sizes. Sometimes, the main microcrack stops growth along the initial direction, and branching occurs. As our researches have shown, it takes place in the cases when (1) its size is ~5 μm and more and (2) the microcrack rests at an obstacle. The next grains, twins, or subgrains can be such obstacles. Electron diffraction, dark field, trace, and stereographic analyses were used. In **Figure 11(a)** micrograph of one interesting microcrack tip area, with the characteristic defects, is shown. As it is apparent from the electron micrograph, the maim microcrack growth along an initial direction is stopped. Ahead of the microcrack in destruction zone, some defects are present. Let us consider them separately.

5.1.1. The main microcrack

TEM study shows that the main microcrack has long, parallel edges along the $<62\bar{4}>$ direction detected in the zone axis [130]. One may suppose that ahead of the crack tip a relaxation of stress occurs that causes the microcrack arrest. However, the relaxation passed through the formation of defects of another type.

5.1.1.1. Disorientation zone

The zone is characterized by dislocation structure, and the respective diffraction pattern corresponds to [130] zone axis. Its area is equal to 15 μm^2, the central part of which is surrounded by peripheral areas. Crystallographic measurements show that the disorientation between the central and peripheral parts reaches 15–16° and both have almost the same dislocation density. In the central part of the disorientation zone, some prolonged dislocations are observed, parallel to the main microcrack. In addition to the above dislocations, some dislocations, angularly oriented to the microcrack, are also observed in the peripheral parts.

5.1.2. Deformation twins

Two parallel twins with the distance between them 0.4 μm are observed in the direction of the microcrack (**Figure 11(b)**, zone axis [130]: upper (a) and lower (b)). The upper twin is two times larger than the lower. The plane of twinning for both twins is $(1^{-}1^{-}1)$, and the direction of the trace is $[62^{-}4]$. The zone axis of the respective electron diffraction pattern is $[51^{-}8]$.

5.1.2.1. Arch-like extinction contour

The detected contour is situated nearly in front of the main microcrack along the microcrack growth direction, between the twins 3 and 5. The contour does not contain the lines of dislocation outcrop. In addition, a dislocation cellular structure is revealed, having the archlike shape in this particular case that may be caused by some non-compensated elastic stresses in the contour. Tilting of the sample causes their widening and narrowing near the tip of the microcrack. Thus, the contours serve as in direct signs of the internal stress localization. However, if such case is observed, the curvature of the contours decreases, and short slip bands and dislocation outcrops are detected.

5.1.2.2. Thin twins

Twins with the twin plane $(1\ 1\ ^{-}1)$, direction of trace $<6^{-}2\ 8>$, matrix zone axis [123], and twin zone axis $[11^{-}\ 2.\ 1^{-}]$.

5.1.3. Short microcrack

The microcrack is oriented at an angle 37° to the main mesocrack and is almost parallel to the thin deformation twins (5). Here, it should be noted that the angle 37° is the angle between the directions $<6^{-}28>$ and $<62^{-}4^{-}>$ and the short microcrack is almost parallel to the deformation twins (5), that is, to the direction $<6^{-}28>$, while the deviation between them amounts to ~5°. The obtained results confirm that in the FCC structures the planes of cleavage, twinning, and slip are the same, while the fracture direction is {111} and is deviated from the twinning shear direction <111> by 6.6°. The above consideration and the analysis of micrographs lead to a conclusion that the destruction zone ahead of the crack tip is confined by the defects (2–6) and these defects practically compose a boundary of the plastic zone. One of the important reasons of steel fracture is a reduction in their relaxation ability during plastic loading. Generally, fracture as well as slip, twinning, fragmentation, disorientation, etc. are the relaxation processes. The fracture (or microcrack propagation) takes place at high stresses, and it is realized after all the above processes are exhausted. Moreover, at high stresses, some processes may flow simultaneously, taking into account the crystallography and Schmidt factor. In the process of stress relaxation in steels, the dislocations will slide along the crystallographic directions (along the planes of slip or twinning) [7].

5.2. Crystallography of the microcracks

Slip bands of different lengths formed near microcracks. The microcracks with long, parallel edges are mainly found inside grains. They are quite long (10–15 μm) and pass through some subgrains, reaching in particular cases the size of some grain diameters. Propagation direction may change within the grain as well while crossing some obstacle such as subgrain boundaries, twins, and carbide precipitates. In addition, to slip bands, near microcracks, stacking faults and dislocation traces are also observed. As it is obvious from the micrographs (**Figure 12**), the microcracks, slip bands, and dislocation traces have the same direction. Moreover, the microcrack propagation direction and the slip bands are always parallel (**Figure 12**). The corresponding diffraction pattern taken from this micro-area shows two FCC zone axes [111] and [114] present simultaneously. The respective stereographic projections show that in the area with the orientation [111] the direction of microcrack propagation is parallel to the slip band. Here, the gnomostereographic projection of the plane (111) is situated in the center of the projection. The microcrack propagation direction and the slip band, in the area with the orientation [114], remain parallel. However, in this case, the normal to the plane of the foil does not lie in the center of the stereographic projection, and consequently the width of the microcrack from the micrograph seems narrower than its real value.

Consequently, the widths of the microcracks in different parts of the subgrain differ. This difference may be derived from the angle between the planes (111) [111] and (11^{-}1) [114]. The results of the calculations are in accordance with those measured experimentally in the micrograph. The performed analyses (statistical, stereographic, and trace) show that the plane of fracture is {111} and the microcrack propagation direction is <110>, in all the considered cases (**Figure 12**). The microcrack propagation direction within subgrain does not vary; however, it changes when passing from one subgrain to another. Thus, the propagation direction is parallel to <110>. The slip bands and stacking faults as well, are oriented along the same directions. This first of all was exhibited in the dislocation structure of the steel. In other words, the stress relaxation occurs just through the formation of these defects. As it was shown by TEM studies, stacking faults, slip bands, and microcracks of two or three different directions were revealed in some grains. The microdiffraction studies and trace analyses showed that the slip occurs along the slip system characteristic of the FCC structure: on {111} planes along the directions <110>. **Figure 12** displays microcracks besides slip bands which are arranged along the slip bands. The microscopic cracks almost always are comparable with the slip bands, both in length and in directions. The boundaries of the micro-area seen in both slip bands and microcracks of three different directions are shown in **Figure 13**. The internal stresses were distributed in the sample and how they relaxed. The trace analysis and microdiffraction studies showed that the direction of propagation of microcracks is <110>. This directly indicates that the microcracks are initiated either in the slip bands or (on) along [8].

5.3. Interaction between microcracks and slip band

As follows from the experimental observations, in practice, only two cases are observed upon the interaction of microcracks with a slip band: when a microcrack propagates in the slip band and when the microcrack is extended at an angle to the slip band. In the first case, the growth of the microcrack will continue until the entire stored energy is spent for its growth or until the microcrack encounters some obstacle after which the microcrack stops. Naturally, during the growth of the microcrack, the possibility of its encounter with another slip bands is not excluded, depending on the level of the accumulated energy. The microcrack will either stop growing along the initial direction or propagate along the direction of the encountered slip band (direction of propagation), or, finally, upon the intersection with the encountered (intersecting) slip band, it initiates a new microcrack into this band, while itself will continue growing along the initial direction of propagation (**Figures 6–13**) [8, 9].

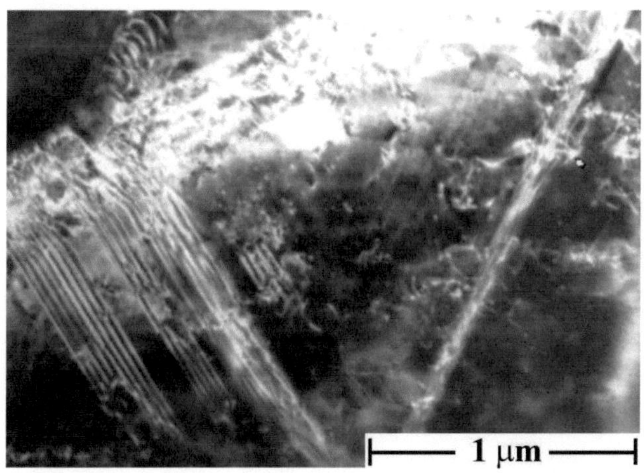

Figure 6. Slip bands and stacking faults. Steel X18 H10T is seen in the upper left part of the micrograph dislocations.

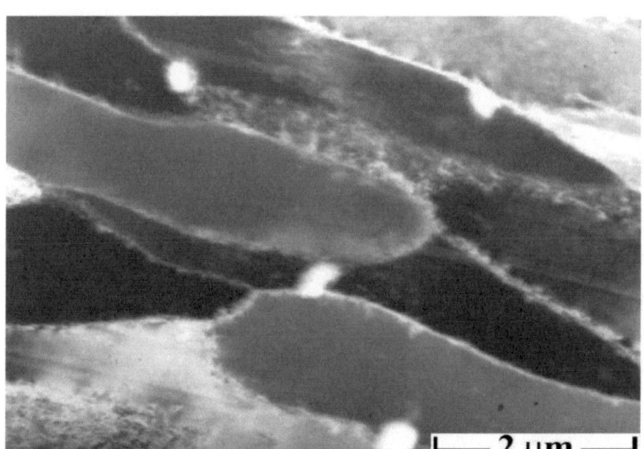

Figure 7. Polygonal structure of the steel after fatigue tests (on the grains of polygons, some carbide precipitates are seen). Direction of polygonization, <110>, steel X18H12Б.

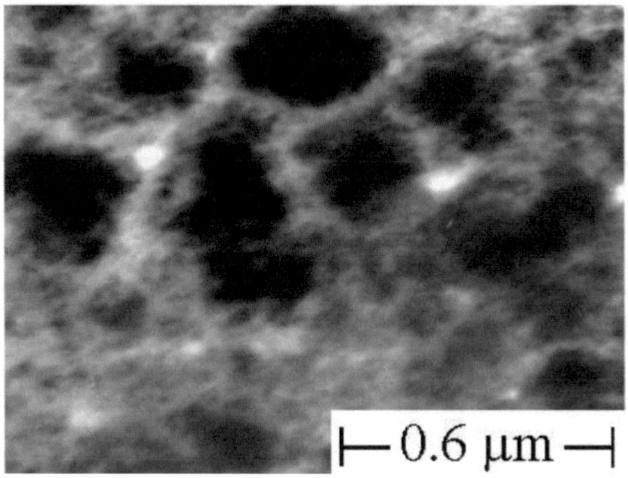

Figure 8. A cellular dislocation structure of the deformed austenitic steel X18H12Б.

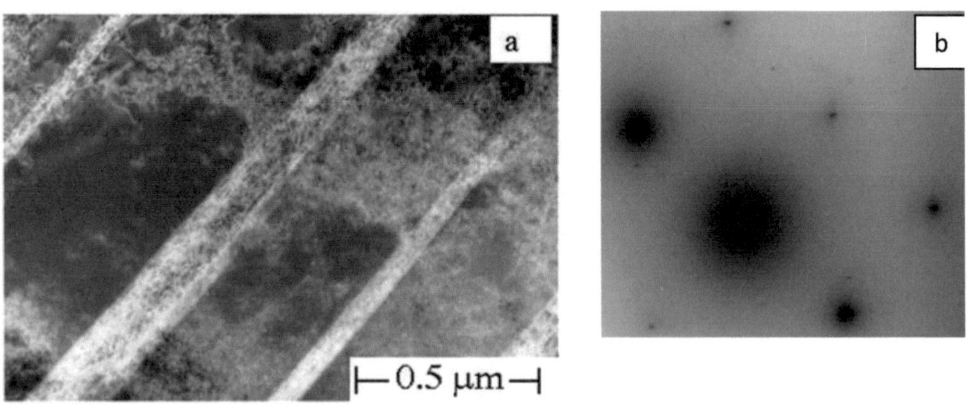

Figure 9. (a) Deformation twins. (b) Zone axis of the matrix [001] is parallel to the zone axis of the twin $[221^{-}]$ (steel X18 H10T).

5.4. Interaction of microcracks with another microcrack

Upon the interaction of a microcrack with another microcrack also, two situations were observed. First, the microcracks propagate along <110> directions as in the same slip band as well as in the adjacent parallel slip bands. In both cases, depending on the level of the stored energy, they can coalesce or stop at a specific distance from each other. However, in some cases, in order to coalesce, they must cause splitting of the area that is located between the microcracks but for this high stresses are required. We did not observe the coalescence of microcracks in this case. In the second situation, when microcracks propagate toward each other at an angle, they almost always unite. In our cases, they stop at small distances from each other. It is likely that in such cases the internal stresses accumulated in the plastic zone are insufficient for the microcracks could continue growing and become coalesced. After the coalescence, the growth continues in

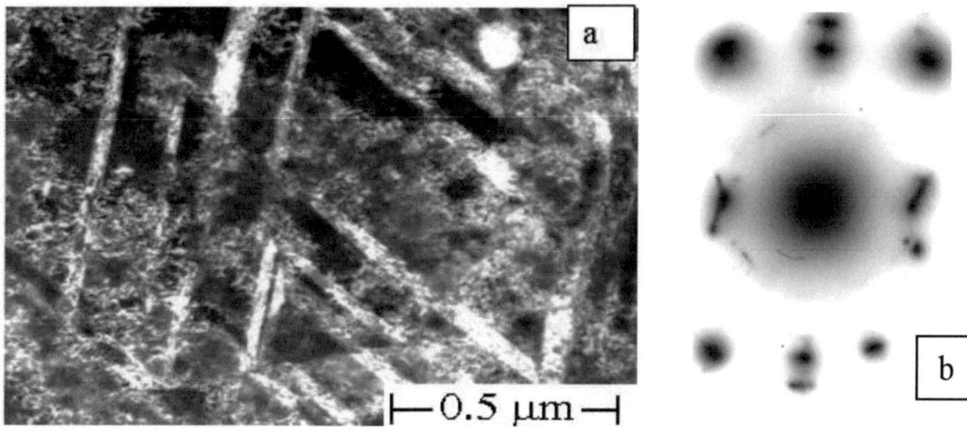

Figure 10. (a) Twins and martensite crystals in the deformed austenite. (b) Zone axis [130] from austenite and reflection (110) from martensite.

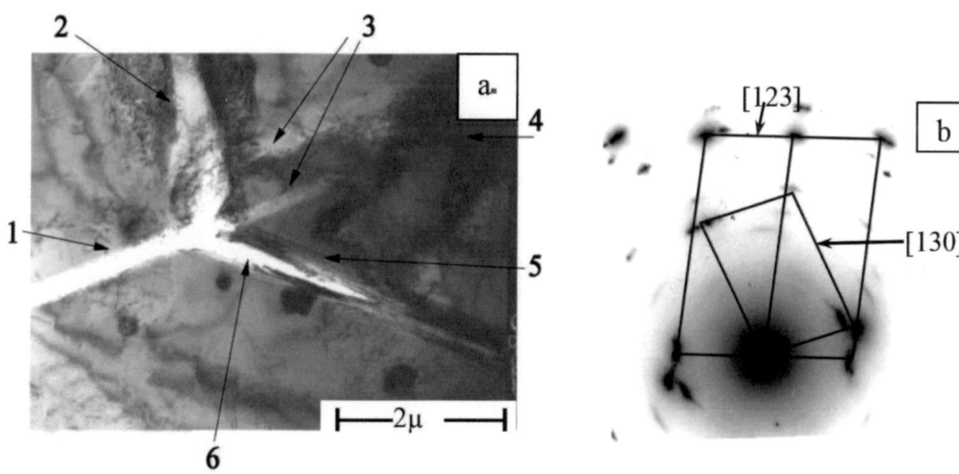

Figure 11. (a) Destruction zone in Cr-Ni-Nb steel. The main microcrack (1); disorientation zone (2); two twins (1) of the direction of microcrack propagation (3); the archlike extinction contour (4); thin twins (5); short microcrack, deviated from the main microcrack but parallel to the twins (6). (b) The indexed electron diffraction pattern, showing traces of twins and microcracks. The scheme is oriented with regard the micrograph.

the direction of one of the microcracks (**Figure 13**). In all the cases examined, it is implied that the direction of the propagation of microcracks remains a crystallographic direction <110>.

5.5. Interaction of a microcrack with carbide inclusions

As it is evident from **Figure 13**, carbide precipitates are also observed in the steel structure together with slip bands and microcracks. In many works it is considered that these are precisely the inclusions of the stress concentrators which are the stacking faults, slip bands, and

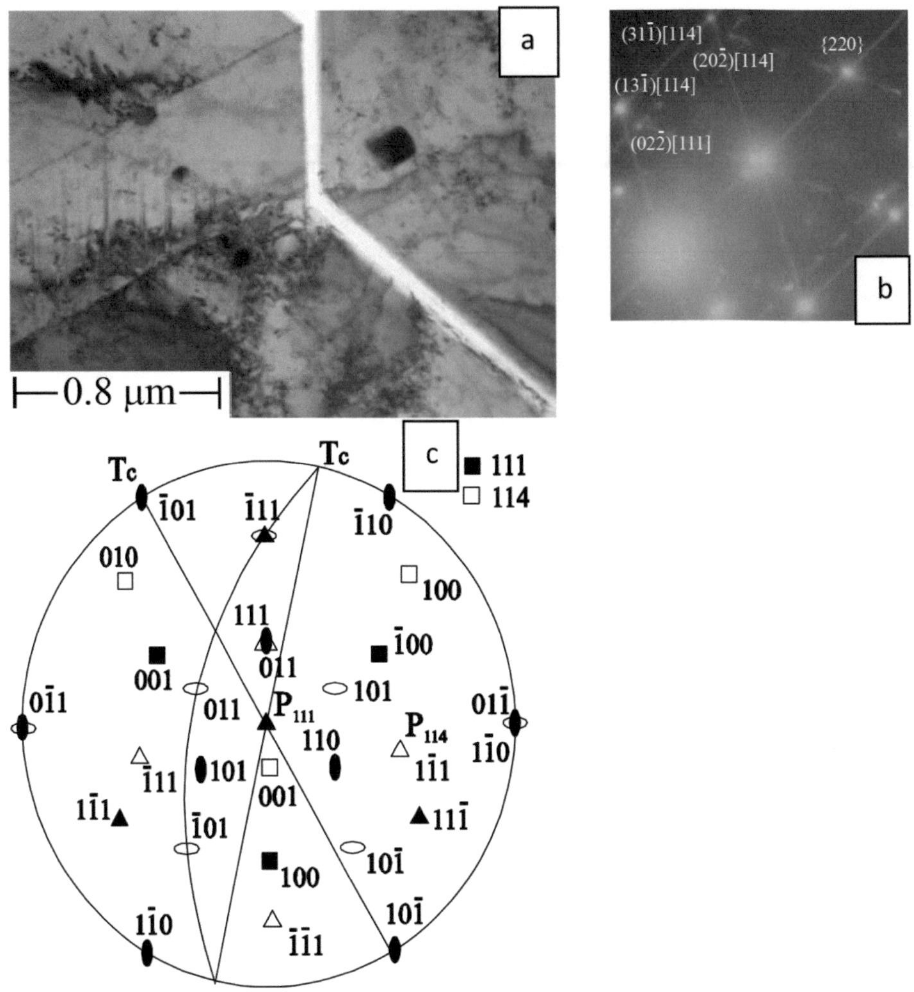

Figure 12. Alteration of microcrack propagation direction at subgrain boundaries, within a grain. (a) A bright-field image of microcrack propagation trajectory (negative image). (b) Electron diffraction pattern. Zone axes [111] (dashed line) and [114] coincide. (c) Stereographic projection.

microcracks in a Cr-Ni-Nb austenitic steel. The electron micrographs show that this is not the case. It is well seen that the inclusions are surrounded by a dislocation tangle (**Figure 13**) and that the dislocation bands ("torches") go out from the region where several inclusions are concentrated. Based on the micrograph, it can be stated that in this specific case the group of inclusions is a barrier to the growth of the microcrack. This follows from the existence of the microcrack whose sharp end rests against the carbide inclusion and the "torches" of stresses are seen apposite to it, which clearly emerge from the tip of the microcrack. They are most likely to be the consequence of the relaxation of stresses that existed before the microcrack tip. Of course, sometimes an opposite picture can be seen, that is, a microcrack that emerges from an inclusion, but this does not make it possible to draw the conclusion that the inclusions

http://dx.doi.org/10.5772/intechopen.70953

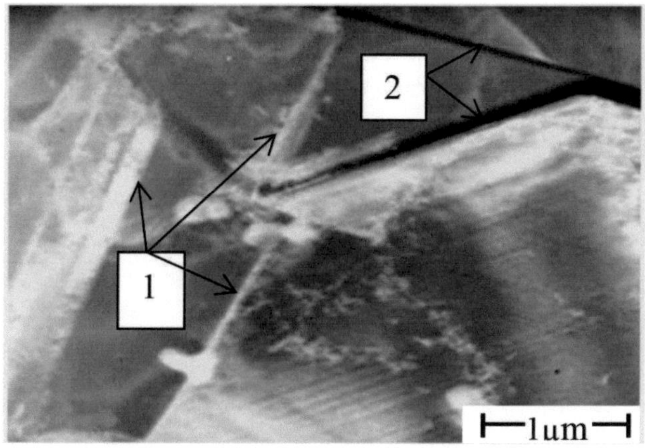

Figure 13. (1) Stacking faults, (2) slip bands and microcracks in a Cr-Ni-Nb austenitic steel.

always initiate the formation of microcracks. The thing is that this process also can depend on factors such as the size of inclusions and the magnitude of elastic stresses before the tip of the microcrack, etc. [9].

6. Discussion

According to the experimental results obtained in the proposed work, a macrocrack is localized within a narrow zone with the width of 600 μm, in which isolated individual mesocracks and slip bands are also observed. To the opinion of the author, the presence of the slip bands in the plastic zone, so that at the initial stage the majority of them are oriented at an angle of 45° to the loading axis and at an angle of 80° after subsequent tests, is not incident. Presumably, when slip takes place at an angle of 45°, at the beginning, it coincides with the direction of maximal tangent stresses. It is obvious to assume that within the plastic zone a local strengthening occurs. After strengthening, another system of slip begins to contribute that coincides or is oriented close to the crack propagation direction. Consequently, the area of maximal deformation shifts to the direction of the main crack. The latter leads to the nucleation of new slip bands, orientated at different angles from the initial ones. This is the evidence of the fact that the slip bands at initial stage of deformation are formed on large microstructure elements. At subsequent deformations slip bands in smaller elements come into action as well [9]. As it was pointed out above, deformation in the plastic zone flows heterogeneously. In addition to the heavily deformed grains, some partially deformed or even non-deformed grains were also observed. In addition, in the locations with no sign of plastic deformation, the isolated microcracks were observed, partly nucleated on (or along) the straight boundaries of grains and microstructure elements of the steel. The distribution and locations of the isolated mesocracks show that the formation of the isolated microcracks is a consequence of stress relaxation. The formation of isolated mesocracks requires at least the following conditions: (1) Grain boundaries should be crystallographic. (2) Slip bands should be oriented parallel to grain boundaries.

(3) Slip should start at grain boundary and then stop. Obviously, the probability of simultaneous implementation of the above conditions is rather small and so is the probability of formation of mesocracks on grain boundaries, especially of their detection. Observation of the isolated mesocracks within the austenite grains (not on the slip bands), in some cases, indicates that the nucleation sites of the isolated mesocracks are the areas of concentration of maximal internal stresses. Thus, during the fatigue tests, demonstration of the following effects is expected: (1) External and internal stresses are summed. (2) External stress facilitates relaxation of internal stresses. Obviously, in the places where the relaxation process is over, the nucleation of microcracks is expected. It is also possible that the relaxation is not finished for some certain reasons; nevertheless, the microcrack is nucleated, but the isolated mesocracks are shorter than the observed slip bands. The latter is obvious from the curve of statistical distribution of the lengths of the isolated mesocracks (**Figures 2–4**). It is clear that there are definite areas and the respective mesocracks with the less length than it was detected in our experiments. It is also difficult because of the simultaneous presence of products of etching, precipitates, grain boundaries, etc. on the polished surface of the sample. Nucleation of the first slip bands on twin and grain boundaries, as well as of the isolated mesocracks, is not a chance. As it is clear from the curves of statistical distribution, the lengths of the slip bands and the microstructure elements are comparable. Therefore, the refinement of grains and the respective microstructure elements of the steel cause the reduction of inhomogeneity of plastic deformation. The number of dislocations in cluster and the amount of accumulated energy are decreasing because the free path of dislocation and consequently the length of slip bands are decreasing. The latter reduces the probability of microcracks' nucleation under the given conditions of fatigue deformation. As it was noted, we observed the macrocracks created after 20–30% of the entire deformation is accomplished. One may conclude that several structural changes occur in the steel just in this period. Finally, some of them contribute to creation of the first microcracks. The deformation flows inhomogeneously as in the plastic zone as well as in the whole sample. From our experiments it follows that in the places of macrocrack deviation, intensive formation of slip bands occurs. As it was shown in the experimental part of the present work, the macrocrack does not preferably passes neither through grain boundaries nor precipitates and does not touch the isolated mesocracks and slip bands as well. Thus, one may suppose that its path, most likely, passes parallel to the elastically stressed microstructural elements of the steel, while the microcrack is nucleated directly in or along the slip bands. It is obvious that when the directions of slip bands are parallel to grain boundaries or to the boundaries of microstructure elements, nucleation of microcracks is simpler. The comparison of the samples undergone to different numbers of cycles but with the same amplitude showed that the increase in frequency directly proportionally decreases the minimal number of cycles leading to the complete destruction. The latter is characteristic to all the investigated steels despite of heat treatment. Hereby, it may be assumed that the fatigue fracture depends on both, the amplitude and the intensity of stress. Accordingly, our investigations showed that fracture in the sample starts with a creation of microcrack. This means that before the formation of mesocrack, several structural changes (slip, twinning) take place in the steel. They initiate creation of martensitic phase and microcracks as well. The microcracks are crystallographically shaped and are nucleated along the slip bands and on grain boundaries, under the favorable circumstances. They are found at a distance of 250–300 μm from the tip of

macrocrack [6]. This fact, once again, indicates that they are of deformation origin and are in close relations with the structural transformations happening in the steel. The revealing of martensite in plastic zone indicates a complexity of mechanism of relaxation of stresses. Apparently, in the beginning, the plastic deformation flows through slip and twinning. However, after strengthening of austenite, martensitic transformation becomes energetically more preferable, or deformation of austenite and martensitic transformation are taking place simultaneously (especially at high values of amplitudes of cyclic deformation)[4]. Consequently, the orientation of micro-cleavages should change while crossing the boundary. The latter is detected experimentally. From our SEM investigations also follow that the macrocrack in steel is created after 20–30% of the entire deformation is accomplished. The latter indicates that at this stage (preliminary), preparatory process takes place in the steel, such as the local and homogeneous slip, creation of dislocation clusters, cells, polygons, etc. Thus, it is generally adopted that mesocracks are formed and propagated on grain and subgrain boundaries. According to our experience, frequently mesocracks and microcracks propagate past subgrain boundaries not touching it. In addition, there are few examples when microcrack passes along the grain or subgrain boundary (**Figures 1**, **2** and **4**). Nevertheless, if the above occurs, such grain or subgrain boundary is strongly linear and even crystallographic, oriented parallel to <110> directions. It is also obvious that in such cases slip bands and dislocation tracks, directions of which are well known for the FCC crystals, are always parallel to the grain boundaries. As what follows from our experimental studies, within 0.5–5.0 hours after preparation of thin foils of the investigated austenitic steel, a redistribution of internal stresses occurs in them. The relaxation and redistribution of stresses occur in homogeneously. In the limits of the examined micro-regions, as was shown above, stacking faults, slip bands, twins, and microcracks are simultaneously observed. Fracture as well as slip, twinning, fragmentation, disorientation, etc. is the relaxation processes, and one of the important reasons of steel fracture is the decrease in their relaxation ability under plastic (dynamic or fatigue) loading. The elastic stresses decrease during fracture, and this leads to the formation of free surfaces. At high stresses, fracture (or microcrack propagation) occurs and is realized after all the above processes are exhausted. However, it does not mean that there is any sequence in the formation of defects. Moreover, some processes at high stresses may flow simultaneously, taking into account the crystallography and Schmidt factor. Thus, the results confirm that ahead of microcrack tip big elastic stresses are concentrated. However, it is quite problematic, and in some cases even impossible, to reveal the field of elastic stresses. In the process of stress relaxation in steels, the dislocations will slide along the crystallographic directions. Therefore, depending on the crystallographic orientation of the thin foil surface, different and nonsymmetric defects will be seen in microscope. Rarely, the archlike extinction contours will also be observed. This means that the observed contour exposes retained elastic stresses existing ahead of microcrack tip. The damaged versions of the proposed model occur in different crystallographic sections (while tilting the specimen), and the scheme projected on the surface of the foil is not always symmetric. As shown in the previous works of the author, the exception is a section through (111) of FCC lattice, taking into consideration the above argumentation. Consequently, determination of plastic zone shape at the tip of a microcrack may be attributed to the theoretical problems (Section 3.4), but for all that, the local crystallography of the material should be taken into account (Section 5). Since slip directions in FCC structures are of

<110> type, with 60° angles between them, and in the BCC of <111> type, with 70° angles between them, the relaxation process in the first case may flow in the six equivalent directions of <110> type, while in the second, just in four directions of <111> type. Taking into account the other ways of relaxation such as twinning in all possible directions, the problem becomes more complicated; in the process of microcrack propagation, the retained stresses are summed; and the relaxation commences through the formation of the above listed defects [10]. On the other hand, the trace analysis and microdiffraction studies inside grain showed that the slip planes are of {111} type and the directions of slip are <110>. The directions of propagation of microcracks are also <110> directions. TEM studies also showed that microcracks are oriented along deformation twins which are crystallographically determined. The microdiffraction, trace, and stereographic analyses of the steel showed that twin plane is a plane of {111} type. The directions of the deformation twins are <62^{-}4 > [130] and <286^{-} > [123]. The directions of meso- and microcracks also are the directions of the deformation twins <62^{-}4 > and <286>. It may be supposed from the above that plane of fracture and the plane of twinning {111} coincide in this case. The directions of microcrack propagation coincide with the directions of traces of twinning plane and the slip bands. Therefore, the crystallography of the material plays an important role in fracture process and propagation of micro- and mesocracks. Namely, crystallography of the material determines the location and nucleation process of microcracks [8, 9]. On the scale of the entire microstructure, the changing crystallographic orientations of grains, naturally, can create inhomogeneous stress fields. The magnitudes of such fields and their influence, on the processes of crack formation, could not be taken into account in the mechanical characteristics upon testing to failure. However, as our studies showed, in the case of small cracks, especially on micro- and nano-level, the local anisotropy cannot be neglected. The formation of slip bands obeys the crystallography of the material. Microcracks are initiated in slip bands and grow along them. The formation and propagation of microcracks are also governed by the crystallography and local anisotropy of micro-regions and the entire grains. Therefore, as a microcrack grows, its shape and trajectory will always depend on the anisotropy of the grains through which it propagates. To this factor, a not uniform distribution of stresses in the bulk of the sample is also added. All this will of course affect the trajectory of the microcrack, in any case, during its growth to meso-dimensions.

7. Conclusions

1. The macrocrack in steel is formed only after 20–30% of the total deformation, and the deformation process in the sample flows inhomogeneously.
2. In the zone of destruction, a martensitic phase is formed, and its volume fraction is directly proportional to the amplitude and frequency of cyclic deformation.
3. The trace, stereographic analysis, and microdiffraction studies proved that the directions of microcracks, slip bands, and deformation twins coincide. The trajectory of microcrack and plane of fracture, within an austenite grain, are governed by the crystallography of the material. A fracture plane is a plane of {111} type and a propagation direction (<110>).

4. The formation of microcracks in slip bands is a consequence of local plastic deformation, whose energy advantageousness is ensured by the relaxation of peak stresses and by the reserve of energy in dislocation cores.
5. Sometimes, inclusions can become barriers for the propagation of microcracks.
6. The study of the interaction of microcracks, with each other, showed that their coalescence or stopping depends on the angle of their joining and on the internal stresses accumulated in the plastic zone ahead of the tips of the microcracks.

Acknowledgements

The research was supported by the ISTC grant G-719, for what the authors are grateful to the Japanese Government, by Shota Rustaveli National Science Foundation, P/N GNSF/ST06/7-072, and by the Georgian Technical University for financing the chapter.

Author details

Tamaz Eterashvili

Address all correspondence to: tetrcsu@gtu.ge

Republic Center for Structure Researches, Georgian Technical University, Tbilisi, Georgia

References

[1] Chang MS, Ritchie RO, Kang YG. Growth behavior of short surface fatigue cracks in 2 1/4 Cr-1 Mo steel. KSME Journal. September 1989;**3**(2):78-85. DOI: 10.1007/BF02953592

[2] Roberts SG. Modeling crack tip plastic zones and brittle-ductile transition. Materials Science and Engineering A. 1997;**234-236**:52-58. DOI: 10.1016/S0921-5093(97)00180-9

[3] Ravichandran KS, Li X. D, Fracture Mechanics. Character of Small Cracks in Polycrystalline Materials: Concept and Numerical K Calculations. Acta Materialia. 2000; **48**:525-540. DOI: 10.1016/S1359-6454(99)00348-1

[4] Eterashvili T, Dzigrashvili T. Study of fracture mechanisms at cyclic fatigue of steels used in nuclear reactors I. Steel Research International. 2012;**83**(3):213-217. DOI: 10.1002/srin.201100184

[5] James R. Dislocation nucleation from a crack tip: An analysis based on the Payer's concept. Journal of the Mechanics and Physics of Solids. 1992;**40**(2):239-271. DOI: doi.org/10.1016/S0022-5096 (05)80012-2

[6] Eterashvili TV. Study of fatigue processes occurring in the plastic zone in front of the tip of a microcrack in austenitic steels (I). Fizika Metallov i Metallovedenie. 2006;**101**(2):200-205. DOI: 10.1134/S0031918X06020128

[7] Eterashvili T. Electron-microscopic study of the formation and distribution of micro-cracks in austenitic steels upon low-cycle fatigue tests. Fizika Metallov i Metallovedenie. 2010;**110**(6):624-627. DOI: 10.1134/S0031918X10120112

[8] Eterashvili T, Vardosanidze M. A fracture crystallography and anisotropy of propagation of microcracks nucleated in stainless austenitic steels after LCF. Key Engineering Materials. 2006;**324-325**:935-938. DOI: 10.4028/www.scientific.net/KEM.324-325.935

[9] Eterashvili T, Dzigrashvili T, Vardosanidze M. Deviations of microcrack during propagation in thin films of austenitic steel and accompanying accommodative processes. Key Engineering Materials. 2015;**627**:297-300, 796. DOI: 10.4028/www.scientific.net/KEM.627.297

[10] Eterashvili T, Dzigrashvili T, Vardosanidze M. The analysis of microcrack tip plastic zone formed in thin films after LCF tests of austenitic steel used in NPP I. Key Engineering Materials. 2010;**417-418**:109-112. DOI: 10.4028/www.scientific.net/KEM.417-418.10